America's Army and Air Force Airplanes

NB
67 060

America's Army and Air Force Airplanes
1918-to the present

Francis H. Dean

Schiffer Military/Aviation History
Atglen, PA

Dedication

This book is dedicated to Evie, Bob, and Donna.

Book Design by Ian Robertson.

Library of Congress Catalog Number: 97-80363

Printed in China.
ISBN: 0-7643-0480-1

Published by Schiffer Publishing Ltd.
4880 Lower Valley Road
Atglen, PA 19310
Phone: (610) 593-1777
FAX: (610) 593-2002
E-mail: Schifferbk@aol.com
Please write for a free catalog.
This book may be purchased from the publisher.
Please include $3.95 postage.
Try your bookstore first.

FOREWORD

The author has accumulated airplane photographs and photo negatives for over 50 years and photographed airplanes for about 20. In the course of collecting he has met a number of fine gentlemen with like interests, many of them members of the American Aviation Historical Society. Much of the photographic material herein is a product of their cooperation and generosity. In addition photographic material has accumulated through 40 years of working in the aviation industry.

Although a large number of US military airplane types and type variants are shown in this book, and the author believes a satisfactory cross-section of airplanes has been illustrated, there is certainly no claim to complete coverage. There are dozens of airplane types not treated, particularly in the early years. But most omissions are fairly obscure and it is believed practically all major types and a high percentage of minor ones are included. The numbers of minor modifications of airplanes are almost infinite in terms of variations in powerplants, airframes, and equipment items.

Rotary wing aircraft have been omitted deliberately on the assumption they deserve separate special treatment.

This book is organized in two sections: major production types with considerable photo caption text, and more minor experimental types with short captions. However with no real excuse a few minor but interesting aircraft have crept into the first section as have research aircraft, and a few production planes sneaked into the second, particularly if a "nice" photo was found. In addition some military planes never purchased by US forces are included as interesting types, particularly where exported in large numbers. The major section's organization is by mission designator as noted in the following list. Short introductions are included in the first section.

MISSION DESIGNATOR	TYPE
A-	ATTACK
AT-	ADVANCED TRAINER
BC-	BASIC COMBAT
B-	BOMBER
BT-	BASIC TRAINER
C-	CARGO
E-	ELECTRONIC WARFARE
F-	PHOTO
P-/F-	PURSUIT/FIGHTER
L-	LIASON
O-	OBSERVATION
PT-	PRIMARY TRAINER
R-	RACER
X- etc.	RESEARCH
T-	TRAINER
U-	UTILITY

A NOTE ABOUT PHOTO CREDITS

Many of the photographs herein are those of the airplane manufacturers; others are from US military sources such as the Army, Army Air Force, and Air Force. These are credited according to the key list below.

Other photos have come from private sources, and these are more difficult to track and properly credit. In many cases of old photos it is impossible to know the original photographer and give him due credit. Aircraft photo collectors and photographers have for years traded private prints and negatives and sometimes not recorded a complete "chain of custody". Where a private photo has been given him the author has credited the donor if no more is known. Where a negative has been given or traded into the author's possession the print from that negative is credited "author" even though he did not take the picture. In fact the author took very few of the pictures in the book. Many were gifts or trades from such long time collectors as Peter Bowers, Roger Besecker, Joe Weathers, Sam Orr, Hal Andrews, and others.

A large number of the earlier photos were obtained by the author in the period 1939 to about 1943 from two sources well remembered by many older collectors. The first was the Craft Art Company in California; these photos are credited to Harry Thorell since it is understood most were from his collection though others may have been involved and some may have been copied from government or manufacturer sources. The second source, also well remembered by many old-timers, was the Aeroplane Photo Supply of Toronto, Canada. This service has not been credited directly since the author's best knowlege is their photos were mainly if not all copies of manufacturer or government shots.

Another complicating factor is that years ago groups of private photographers would haunt easily accessible airports of those days and often take essentially the same views of a new aircraft that had just come in. In some instances they would take several identical views to have trading material for use with other collectors around the country. These are some of the factors making it often difficult to assign proper credit for a private photo. The author has done his best to give an honest accounting of photo crediting. A list of the credit lines follows.

Photos are credited to the manufacturer with exceptions as follows:
NATIONAL ARCHIVES 187
AUTHOR 8,9,12,13,21-23,29,30,A-2A,A-2B,A-3D,A-11A,37-39,45,49,51,57-61,64-69,71,72,76,78,79,82,85-92,95,97,99-101,A-17B,A-18B,A-18D,A-32A,A-33,A-38A,A-38B, A-40B,103,107-109,119,123,128,130,132,133,140-144,146-148,150,155,158-160,164,165,A-52D,A-55B,A-64D, A-64G,A-64J,A-65A,A-68A,A-68B,A-74,177-182,184,186,A-79A,A-80A,196,201-204,208-210,212-215,218,222,224, A-81B,A-81C,A-89A,A-90A,A-91A,A-91B,A-93A,A-94A,A-97A,226-228,232,236,237,240-242,245,247,248,251-254,257-259,261,267,269,271,272,274-276, 277, 278, 280, 284, 286, 288,289,293,294,297-299,301-304,308-314,316,317,319-322,324,325,328,330,332-335,338,341-343, 345, 346, 348, 350, A-105A,A-125B,A-127C,A-129,A-129C,A-148, A-151,A-154C,A-155,A-156J,A-160,A-163,A-163A,A-164A, 352,357,359,360,366,A-171A,A-173A,A-174a,373-378, 380, 383, 385,386-390,A-175C, 391, 392, 393, 395, 397, 398A, 400, 403,407,408.
ROGER BESECKER 5,25,A-10,70,A-21,A-48B,A-60A-64,A-65B, A-66,A-72,A-73,A-76,A-88,A-92,260,326,A-113D,A-156A,A-165,363
P.M.BOWERS A-7
DON CAMPBELL A-68
MARTIN COPP A-82A,A-129D
CURTISS PROPELLER DIV. A-149A
WILLIAM DUNTON A-128
RENE LA CLAIR A-16,250
BIRCH MATTHEWS A-137,A-138,A-139
HARRY MUTTER A-72A
NACA A-146
R.L.STUCKEY 77,156,183
HARRY THORELL 1,A-3B,A-3C,A-3a,A-5,A-6,A-6A,A-8,A-11,50,A-19,A-20,A-22,A-23,A-24.A-25,A-41,104,105,A-44, A-44A,A-44D,A-45,A-46,A-47,A-48,110,112,112A,113-118, 120,121,A-49,A-50,A-50A,A-51,A-52,A-52A,A-52B,A-53,A-54,A-55,A-55A,A-56,A-57,A-58A,A-59,A-61,A-62,A-63, 169,170,A-70,192,193,194,200,205-207,216,220,221,A-81,A-83,A-84,A-86,A-89,A-90,A-91,A-94,A-95,A-96,A-97,A-104, A-104A,244,246,249,A-106,A-126,A-156C,355,358,361,A-168,A-170,A-171
US ARMY 19,A-0,43,46,161,175,176,A-79,195,A-81D,A-86A, 229, 239,A-104B,A-104C,A-104D,A-108,A-108A,A-113A,A-120A,A-124A,A-127B,A-129B,A-131A,A-132,A-133,A-145A, 354,A-167,369.
US ARMY AIR FORCE 11,15,24,75,134,285,A-134,A-141A,A-152C.
US AIR FORCE 17,63,84,145,149,305,347,A-150A,A-156B,A-156F,A-156I,A-172,379.
US NAVY 16.
WILLIAM VANCE A-88A
JOE WEATHERS 362,A-175B.
WRIGHT AERONAUTICAL CORP. A-48D
(UNKNOWN) A-12,A-18C,A-18F,A-27,A-43, 111, 138, 153, 154, 166, A-77,A-82,243,256,307,A-112A,A-131B,A-136,401.

INTRODUCTION

This book provides an overview of the progress of American military airplane design from approximately the end of World War I in 1918 to the present by means of extensively captioned photographs of the major types along with an additional picture gallery of more minor, mostly experimental aircraft in a supplement. It is organized in alphabetical ordered sections according to the basic mission designators of the aircraft. A period of slightly over three quarters of a century of aviation progress is thus covered, and advances have been truly extraordinary. Military airplane high speeds have increased from under 100 miles per hour to ten or twenty times that figure. Altitude performance has been extended from no more than a few thousand feet to the limits of the earth's atmosphere, and range capabilities from perhaps 100 to many thousands of miles. Airplane design configurations have progressed from drag-laden open cockpit tail-dragger biplanes with multi-bay strut and wire-braced wings and fixed landing gear to sleek low drag enclosed cockpit cantilever monoplanes with flush-retracting landing gear.

Aerodynamic advances include new low drag wing airfoil shapes and specially contoured area ruled fuselages. Structural and materials advances have been many. Fabric covered wooden framed structures gave way first to wire braced metal tubular or square section welded or bolted truss framing under fabric, or sometimes corrugated sheet metal covering, and later semi-monococque stressed skin all metal construction and lately use of sandwich structures for great stiffness and composite materials. Special heat resistant materials are employed to withstand aerodynamic heating of airframes at very high speeds and the local effects of gas turbine engine exhausts.

Powerplants have undergone a similar revolution, progressing in the piston engine field from low power-to-weight water cooled engines of uncertain reliability on to high power multi-cylinder air or liquid cooled models of various configurations with a high degree of supercharging or compounding. Power-to-weight performance continuously improved as did specific fuel consumption figures. During the World War II period the aircraft gas turbine became a practical reality sired by the impetus of wartime, and the jet, turbopropeller, and turbofan engines derived therefrom became realistic in the latter portion of the period and a revolution in airplane design followed.

In the subsystems and equipment areas similar improvements in technology have continued apace. Instruments and displays have changed from a very few large heavy awkward items to compact multi-function units of great variety and versatility, including cathode ray tube and heads-up displays. Electric and hydraulic utility subsystems have gone from non-existance to highly efficient multi-channel use. Flight controls have improved from cumbersome all-mechanical arrangements to multiple path fly-by-wire and fly-by-light systems using power-boosted control surface actuation. Perhaps the greatest advances of all have been made through the period in avionics, or aviation electronics. Starting with possible installation of a single large heavy simple radio in the early periods, the capabilities of aircraft avionics have increased to a startling degree in areas of communications, navigation, identification, weapons control, and countermeasures along with integration of these functions in a manner allowing reasonable crew workloads. Photographic, radar, and infrared sensors provide continuous data input to various systems.

Aircraft weaponry has changed dramatically from installation in combat machines of a single rifle caliber machine gun through multiple heavy machine guns and cannon, sometimes in special remote radar-directed turrets, along with rockets and bombs of almost infinite design variety, including remotely controlled "smart" bombs and rockets.

The problems along the way have been many, and solutions have often been purchased at the cost of human lives in the flight test areas, both on the part of contractor and military service flight personnel. As the complexity of military aircraft has increased many-fold through the years covered herein the problems of achieving proper reliability and good maintainability have been constant challenges and remain so today. The tremendous cost of the modern military airplane is a problem being constantly faced, particularly when peacetime military budgets must be severely restricted. In years well past many aircraft companies could produce a new prototype military plane at least every other year, and several companies could thus compete for production contracts with prototype hardware. Today just putting together a paper proposal against military rquirements is a wrenching event for even the largest of companies, and a combination of effort by company teaming is almost inevitable. Evaluating the paper proposal by military personnel is also a mind-numbing endeavor. Supplying a hardware prototype via company funding alone is now seldom possible.

The number of aircraft types is reducing and their longevity is of necessity increasing. Cases of military airplanes being older than their crews are now a reality.

It is hoped the reader will gain some understanding of historical trends for US military airplanes over the period in an easy and entertaining manner by viewing the photos and reading the accompanying captions covering about three quarters of the current 20th century, and wonder what the next century will bring. And some long-time aviation enthusiasts may even spot some aircraft they have not laid eyes on before, or a new view of one or two!

ATTACK AIRPLANES

For the Army the attack airplane was a tactical weapon intended to destroy opposing troops and enemy installations in direct support of the operations of friendly forces. Generally there were two methods of attacking the enemy. The first was a low level flying assault using ground strafing fire from the guns of the attack plane plus loosing relatively light anti-personnel bombs upon the targets, normally employing a large number of aircraft to "smother" the area defenses. The second was a vertical or near vertical dive bombing attack aiming the planes at their targets. The dive would start at a medium altitude with bomb or bombs release and pullout and escape at low altitude. Army planes for both means of attack were procured, but those made specifically for dive bombing had to wait until the World War II era. The earliest US Army planes for attack missions tended to be heavily armed and armored "aerial tanks" designed to slug it out nose to nose with defense installations. Slow, large, and ungainly, these ground attack aircraft such as the Engineering Division GAX and Boeing GA-1 and -2 of 1919-1922 were soon, and luckily, perceived to be a wrong answer.

In the late 1920s the Army turned, on an interim basis, to modified two place observation planes with extra forward firing guns but no armor as attack planes. For the next decade Army attack planes followed the same general formula, changing only with aircraft technology changes-biplanes turned into low wing monoplanes, cockpits were enclosed, and the planes got cleaner and faster. Armament changes were small, four to six fixed forward firing rifle caliber machine guns plus a flexible gun or two for protection to the rear along with a light load, perhaps up to 600 pounds, of bombs. The only real tests such planes got in actual combat were the experiences of the Northrops and Vultees exported to China to fight the Japanese in the 1930s. It is not clear that these planes had much tactical success. Similar aircraft, the British RAF Fairey Battles, took a terrible drubbing from German fighters very early in World War II. A change in attack plane design, heralded first by the faster Curtiss twin-engined XA-14 and A-18, was nailed down by the Douglas A-20 Havoc, and later by its successor, the A-26 Invader, really more light bombers, but with multi-role capability. The Havoc could not dive bomb, however, and the early success of the German Stuka dive bombers convinced Army planners they needed a tactical dive bomber for close air support. During World War II they procured two Navy dive bomber designs, 1700 plus A-24 and A-25 aircraft, neither used operationally to any extent, 500 A-36s, a converted early Mustang fighter of limited bombing capability, and one true dive bomber, the Vultee A-31/A-35, used only by the British in its intended role. For the attack mission in World War II the USAAF employed fighter aircraft with external bomb-carrying capability, noteably the P-47 Thunderbolt.

After the war in 1948 the Air Force dropped the A for attack designation and redesignated their only attack airplane, the A-26 Invader, as te B-26, used effectively in the Korean War. There were no more Air Force "attack" airplanes for some years, not until early in the Vietnam War when the USAF again found itself without a dedicated ground support aircraft, a role they never relished anyway. So again they obtained a Navy design, the A-1E Skyraider for use in early Vietnam operations. The other aircraft utilized by the USAF in Vietnam for ground attack were conversions- the A-37 Dragonfly counter-insurgency aircraft was a conversion of a T-37 trainer, the OV-10s were not so much attack aircraft as forward air controller types, the A-7D of 1968 was a revised Navy airplane utilized by the USAF for ground attack in the latter stages of the conflict, and the AC-130, a cargo plane conversion, was another "attack" plane of sorts. In the late 1960s the USAF, again pressured by the Army to procure a true attack airplane for ground support (the Army was not allowed to procure one of their own), sponsored a competition for an AX tank-killer airplane. A fly-off in 1972 between the Northrop A-9 and Fairchild Republic A-10 resulted in a win for the A-10 and a production contract. Today the A-10 Warthog, though out of production by a now non-existant company, is the Air Force attack airplane, and was employed with success in the Gulf War. What the future holds for the attack class aircraft is questionable. The USAF loves "the blue sky yonder", and in recent years has seemed less than enthusiastic about the Army ground support mission, though current USAF fighters have ground attack capability.

1. 221CURTISS A-3B The Falcon, shown here in the A-3B version, was the Army's first real attack airplane, a modification of the O-1B and O-1E observation types. Powered by a 435 horsepower water-cooled Curtiss D-12 engine, this metal-framed fabric-covered two place biplane carried four .30 caliber forward-firing machine guns, two .30 caliber guns in the rear cockpit, and 200 pounds of bombs underwing. A total of 154 attack Falcons were delivered from October, 1927 into 1930, equipping four ground attack squadrons. Reliable 140 mph maximum speed airplanes, some lasted as late as 1937 as utility hacks.

CURTISS A-3B A company photo shows an updated arrangement of an A-3B attack airplane front cockpit and instrument panel, providing a fine view of a 1930 military plane front office. Engine instruments are at upper center, flight instruments on right and lower center panels, and engine throttle and mixture controls in the standard left side position. Butts of the two synchronized machine guns are readily accessable.

3.Curtiss A-8 An A-8 Shrike, fourteen of which were built as prototype and service test aircraft for the Army in 1932-3 after Curtiss won a 1931 competition for a new two-place monoplane attack aircraft. Innovations included the thin wire- and strut-braced metal monoplane wing with flaps and leading edge slats, and enclosed cockpits for the metal fuselage. A 600 horsepower liquid-cooled Curtiss Conqueror engine provided a high speed of about 185 mph. Four .30 caliber guns were in the fixed wheel fairings; the rear gunner had a .30 caliber weapon, and up to 400 pounds of bombs could be carried.

4.CURTISS Y1A-8A Flight photo showing a November, 1932 view of a Shrike using a revised Conqueror engine with propeller reduction gearing, the noise of which resulted in the airplane being called "the old coffee grinder". Note the separated cockpits, external fuel tank, modified nose contours, and the increased protrusion of wheels from their fairings as shocks extended in flight. Shrikes were also proposed by Curtiss as two-seat Army fighters and as Navy catapult seaplane scout and observation planes with radial engines. There were no takers.

5.CURTISS A-12 A "production" version of the Shrike runs up its 690 horsepower nine cylinder radial Wright Cyclone engine, a major change from the liquid-cooled engine of the A-8. The air-cooled installation reduced vulnerability of the Army ground attack airplane but also reduced speed slightly. The A-12 was otherwise generally similar to the A-8 with four landing gear mounted and one rear .30 caliber machine guns. Bombs or an external fuel tank could be carried under the fuselage between the landing gear legs.

6.CURTISS A-12 Flight view of an A-12 taken by the company photographer shows large frontal area of the Cyclone radial engine and open front cockpit. Pilots were apt to be suspicious of closed canopies in those days. Some later A-12 aircraft had a raised turnover structure to protect the crew. The A-12 airplanes started delivery in late 1933, most all going to the Third Attack group at Fort Crockett, Texas, and operated "front line" until 1936. Ten additional export Shrikes got to China.

7.NORTHROP 2-E Though Northrop had no success initially in selling their new two-place all-metal low wing Model 2 attack plane to the US Army, it did well in the export market. The 2-E Gamma type shown here in February, 1934 with Wright Cyclone radial engine, full cantilever wing, fixed landing gear, and fully enclosed cockpits went to China in quantity, with 49 aircraft shipped and used against the Japanese from 1934 to 1938. High speed was 219 mph. Armament comprised two fixed wing and one flexible rear .30 caliber guns and a bomb load up to 1000 pounds carried in racks between the landing gears. One 2-E each went to England and Russia.

8.NORTHROP A-17 After a long period of wooing the US Army with their attack design Northrop finally received, as a 1934 Christmas present, a two million dollar order for 110 A-17 airplanes considerably modified from the 2-E in using a twin row Wasp Junior radial engine of 750 horsepower and revised cockpit canopy and landing gear. Additional changes involved four wing guns and a rear flexible weapon along with internal stowage for twenty thirty-pound bombs in chutes. Production deliveries started early in 1936.

9.NORTHROP A-17A An A-17A taxis out for takeoff. This version of the A-17 featured retractable landing gear with wheels folding into wing root pockets along with an uprated Twin Wasp Jr. engine of 825 horsepower. These changes improved high speed about 15 mph to 220 mph. A total of 129 attack types plus two three-place command transports were ordered by the Army with deliveries starting in January, 1937 and ending late in 1938. These airplanes replaced the A-17s in attack squadrons. In mid-1940 most A-17As were taken back by Douglas (which had taken over Northrop) to be sold overseas to England and France.

12.VULTEE V-11 Another mid-1930s attack bomber, this by Vultee, is shown in side view. Characterized by a Wright Cyclone radial engine of 750 to 850 horsepower, sleek lines, a cantilever wing mounting retractable landing gear, and a long "greenhouse" canopy, the three-place Vultee had both dorsal and ventral rear gun positions and four .30 caliber wing guns. Internal stowage for twenty thirty-pound or external provisions for up to 1250 pounds of bombs was designed in. After first appearing in late 1935 a total of 102 V-11s were exported, including 40 to Turkey, 29 to China, and 25 to Brazil. A few went to Russia. Exports of this attack plane were made through 1939.

13.A run-up picture of the Vultee attack bomber shows the thick all-metal cantilever wing of 50 foot span and the pockets for retractable landing gear. High speed was about 230 mph and normal loaded weight was just under 10000 pounds. The third man operated either as a prone bombardier or ventral gunner. The aircraft had both fuselage and wing fuel tanks for a total of almost 500 gallons. The US Army ordered only seven Vultees as the YA-19 with Twin Wasp engines. Some advanced V-12 versions were sold to China.

10,CURTISS A-18 This first venture into a twin engine two place Army attack airplane is shown in a Curtiss photograph of summer, 1937. Based on the design of an earlier 1935 XA-14 prototype, the trim all metal A-18 used two Wright Cyclone radials each of 930 horsepower, and achieved a 240 mph maximum speed. A July, 1936 order for thirteen service test airplanes was completed by late 1937, costing just over $100,000 each. A-18s operated with the Third Attack Group, later the Third Bombardment Group, as training aircraft.

11.CURTISS A-18 A flight view of the A-18 shows off a slim two place fuselage and two large diameter Cyclone radial engines out ahead. Armament comprised four .30 caliber nose and one aft .30 caliber weapon. Internal bays carried up to 650 pounds of bombs. The A-18 wing structure was all metal in three sections with center section having three spars and outer sections monospar. Much of the wing was fabric covered. Slotted flaps were included. Retractable landing gear was operated by an electric motor powered hydraulic system.

14.DOUGLAS DB-7B An export DB-7B Boston destined for Britain is seen flying over Los Angeles. A limited number of earlier DB-7s got to France in 1940. This very useful light attack bomber type served throughout World War II in many versions. Except for early French models, these aircraft were powered by two Wright two-row fourteen cylinder Double Cyclone engines each of 1600 horsepower providing a high speed of over 300 mph at medium altitudes. A bomb load of 2000 pounds could be carried. More than 800 DB-7 types were manufactured, 240 of them by Boeing.

15.DOUGLAS A-20A An early A-20A light attack bomber, the US Army version of the export DB-7 is shown in flight. The A-20A was the first production Havoc, a total of 143 being produced. With a gross weight of ten tons and a high speed of 347 mph the A-20A had seven machine guns, four in the nose, two dorsal, and one ventral. The type first flew in the fall of 1940; all were delivered the next year. Later models had .50 caliber guns.

16.DOUGLAS A-20A This A-20A photo of January, 1941 shows one of the nose gun packs housing two .30 caliber weapons. Actually this airplane is the single Army A-20A supplied to the Navy for test as a BD-1 in December, 1940. Other A-20As went to the Army's light bombardment groups. The Havoc had a crew of three with a single pilot in its narrow fuselage. The A-20 series was to become the standard USAAF World War II light bomber.

17.DOUGLAS A-20G. The G model Havoc was put out in greater quantity than other versions with a total production run of 2850 airplanes. It incorporated a solid gun nose using either four 20mm cannon or six .50 caliber machine guns, a twin .50 caliber gun turret aft in the dorsal position, and a single .50 caliber ventral gun. By this time normal gross weight had increased to over twelve tons and high speed reduced to about 315 mph. Almost 6300 Havocs were produced; about 3000 of them went to the Russians.

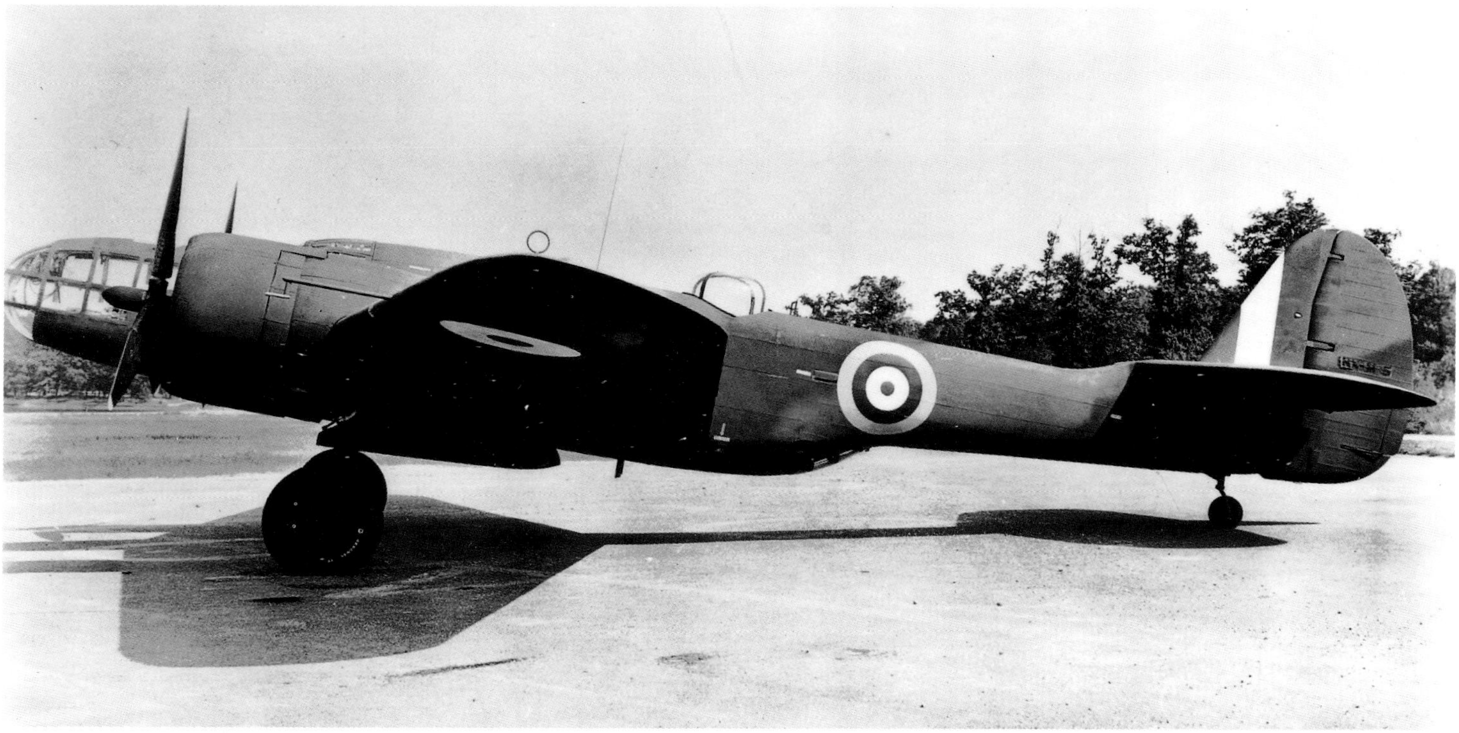

18.MARTIN 167 This export light attack bomber, labeled the Maryland by the British, was a loser in the 1938 US Army attack aircraft competition, but was produced in quantity for the French and British. The French operated some against the 1940 German invasion and the British and South Africans used it in North Africa. A clean three place moniplane with two Pratt and Whitney Twin Wasp engines of 1050 horsepower each, the 167 achieved a high speed of just over 300 mph and had exceptional range. Over 300 Marylands were built.

19.DOUGLAS A-24 The Army equivalent of the famous Navy SBD- Dauntless dive bomber, the A-24 did not achieve similar success. The few used against the Japanese during early 1942 in the Southwest Pacific were mauled by enemy aircraft unless heavily escorted by friendly fighters. A total of 863 A-24 dive bombers were produced for the Army. They used a Wright Cyclone of 1200 horsepower which gave a high speed of about 250 mph. Armament included two .50 caliber cowl and two .30 caliber rear flexible guns and a maximum bomb load of 1200 pounds.

20.CURTISS A-25 The A-25 was an Army version of the naval SB2C-1 Helldiver, the airplane shown being an A-25 intended for export to Australia. Though 900 A-25 dive bombers were produced from late 1942 to early 1944 at the Curtiss Wright plant in St.Louis, Missouri, none went into combat. Ten were destined for Australia; the Army used others as trainers and eventually turned about 400 over to the Marines, these being labeled SB2C-1As. Powered by a fourteen cylinder Wright Double Cyclone engine of 1700 horsepower, the A-25 achieved a high speed of about 280 mph and weighed about eight tons. A 2000 pound load of bombs could be carried internally. Armament was composed of five .50 caliber guns, four of them fixed in the wings and one in the rear.

21.DOUGLAS A-26B Designed as a successor to the A-20, the lean long-lived Invader light attack bomber first flew in mid-1942. Shown is the B attack version with solid gun nose having six or eight .50 caliber weapons. Twin gun upper and lower remote turrets were incorporated. Rocket mounts are shown under the wings. Two tons of bombs could be carried internally, and two Pratt and Whitney Double Wasp engines totaling 4200 horsepower made the A-26B a 350 to 370 mph airplane. A total of 1355 attack versions were produced in two factories. The type was in European combat starting in late 1944, was used extensively in Korea, also in French Indo-China, and later in Vietnam. In 1948 the attack class was dropped and the A-26 became the B-26 bomber.

22.DOUGLAS A-26C Shown is the A-26C version with transparent bombardier nose. This model was used as a lead aircraft for bombing in Europe late in World War II. It normally had two fixed nose guns with other armament like the B version. It had an internal bay for 4000 pounds of bombs and provision for wing shackles carrying up to an additional 2000 pounds. A total of 1091 C versions came off the production lines during World War II. In clean configuration the A-26C could hit 370 mph. Like the B airplanes, A-26C/B-26C bombers had long lives, serving in Korea and initially in Vietnam.

23.DOUGLAS-ON MARK YB-26K The prototype of a re-manufactured B-26 by On-Mark Engineering, made necessary by structural failure of B-26 aircraft early in the Vietnam War. First flown in the spring of 1964, the YB-26K was successful enough to merit an order for 40 more aircraft conversions. The K model was easily distinguished by its use of wingtip tanks. Powered with 2500 horsepower uprated Double Wasp engines, the B-26K had a gross weight of almost 20 tons and a high speed of 323 mph. A maximum of 12000 pounds could be carried internally and under the wings. The type saw limited use during the Vietnam War until 1970.

24.LOCKHEED A-29 The Lockheed Army A-29 Hudson shown was an attack airplane only in the sense that it was often used to quite successfully attack submarines. Ordered for coastal patrol by the British in mid-1938, the Hudson was a navalized version of the commercial Lockheed 14 Super Electra transport. Over 2600 were built in several versions; the A-29 models employed Wright Cyclones and the similar A-28 types used Pratt and Whitney Twin Wasp engines, both in the 1000 to 1200 horsepower class. Some Hudsons had a dorsal power turret; others had an open cockpit in that location. They carried an internal bomb load of 1200 pounds and had a maximum speed of about 250 mph. Late versions employed under-wing rockets.

25.MARTIN A-30A The follow-on Martin attack bomber to the Model 167 Maryland, the 187 Baltimore was known as the A-30A by the USAAF though never used operationally by American forces. The Baltimore was designed to meet French and British requirements in 1940 with Britain taking over French orders under Lend-Lease. A total of six versions were produced with changes mainly in armament. The A-30A (Baltimore V) shown had two Wright GR-2600 Double Cyclones of 1700 horsepower each providing a high speed of almost 320 mph. This light bomber version had armament of four .50 caliber wing guns, a two .50 caliber gun Martin dorsal turret, and one flexible .50 caliber gun in a ventral position. It could fly over 900 miles with 2000 pounds of bombs internally. A total of 1175 Baltimores were produced.

26.1DOUGLAS A-33 Export versions of the US Army A-17 and A-17A, the Douglas 8A series culminated in the 8A-5, later the Army A-33 attack bomber. Previous 8A versions had been the 8A-1 (one to Sweden), the 8A-2 (30 to Argentina), 8A-3N (18 to the Netherlands), 8A-3P (ten to Peru), 8A-4 (15 to Iraq), all exported from 1938 to 1940. A final version, the 8A-5 shown here, was for Norway, but because of the Nazi takeover the 36 planes with a 1200 horsepower Wright Cyclone engine and a high speed of about 250 mph went to "Little Norway" in Canada. In 1942 these planes, less training losses, were taken over by the US as A-33s. They were never used in the light attack role.

27.Vultee A-35 A true army dive bomber, the A-35 Vengeance was a follow-on to the export V-72 to Britain and the earlier A-31 dive bombers. An angular two-seater using a Wright fourteen cylinder Cyclone engine of 1700 horsepower providing a maximum speed of about 275 mph, the A-35 had five to seven .50 caliber guns with a maximum of 2500 pounds of bombs internally and the same externally. The USAAF did not use the Vengeance operationally, preferring bomb-laden fighters, but the British used the type in the Far East, principally in the Burma campaigns.

28.NORTH AMERICAN A-36A The A-36A shown here as a boneyard queen was an Allison engine powered Mustang fighter fitted out with dive flaps and bomb racks to act as a dive bomber. Armament consisted of six .50 caliber guns and two 500 pound bombs, one under each wing. Starting in the fall of 1942 a total of 500 A-36A airplanes were produced. In mid-1943 two USAAF groups used the A-36A in the Mediterranean area with some success; they were replaced by less vulnerable Thunderbolts the next year.

29.CESSNA A-37B An A-37B Dragonfly loaded with a Minigun in the nose section, tip tanks, refueling probe, and plenty of external stores on eight wing pylons shows off for the company cameraman. An attack version of the Cessna T-37 twin jet advanced trainer, the A-37B was a follow-on to 39 A-37A airplanes built in 1967. Almost 600 A-37s were produced; many were distributed to foreign governments. Powered with two General Electric J85 jets of 2850 pounds thrust each, the 500 mph A-37B served into the 1980s.

30.Vought A-7D An Air Force version of the Navy A-7 Corsair II, the A-7D was to be used for close support and save the cost of developing a completely new attack airplane. Changes from the Navy version included a new Allison turbofan engine, new armament, and different avionics. The A-7D went into service in 1970 and operated in Vietnam with a good record until the bitter end. With a gross weight of about 20 tons and a high speed of over 650 mph, the Corsair II had a 20mm Vulcan fast-firing cannon and could carry over 6000 pounds of bombs.

31.FAIRCHILD-REPUBLIC A-10A Known during the 1970 design competition as the AX (Attack Experimental), the A-10A won a fly-off with the Northrop A-9A design to go into production in 1975 as an Air Force twin turbofan-powered ground attack plane. Known as the Warthog for its unhandsome looks, the large and relatively slow A-10A has a high firing rate 30mm cannon in the nose and multiple store stations underwing for carriage of ground attack missiles or bombs. It is also heavily armored and, with a big wing, quite maneuverable. A-10As were used successfully during the recent Gulf War. Powered by two 9065 pound thrust General Electric TF34 turbofans and with a maximum gross weight up to 25 tons and wingspan of 57 feet, the A-10A has a high speed of about 425 mph.

ADVANCED TRAINERS

The advanced trainer (AT-) category of airplanes in the Army was instituted in May of 1924 and lasted well past World War II, though the last AT designated type was converted in 1944 from a bomber. Some time after that war's end in 1948 the various trainer categories, including AT, were all put into a simple T for trainer category with no further distinction, and existing types in service retained their previous AT number-thus an AT-6 became just T-6.

In the 1920s the advanced trainer series started with a few Consolidated, Dayton-Wright, and Huff-Daland biplanes, some of them with side-by-side seating. On the basis that the advanced trainer was the last type the student pilot would fly before taking over a high performance combat-type aircraft, in 1926 the Army decided to use de-rated fighters as advanced trainers. Versions of the Boeing PW-9 and Curtiss P-1 Hawk pursuits were designated as AT-3, AT-4, and AT-5 trainers with Wright Hispano engines of less than half the power installed in the fighters. The experiment was not a success, and within two years the trainers were re-converted with fighter engines.

The first "modern" advanced trainer was the AT-6, the North American Texan, with thousands built starting in 1940-41. At that time the student pilot, having completed basic training, got onto the two place AT-6 with a 600 horsepower engine, retractable landing gear, and more instruments and radio, including a radio range beacon receiver. Emphasis was on cross-country flying, both day and night, instrument flying and orientation problems with radio direction aids, night navigation and increased formation flying. The working day included about two hours of flying and a lot more in the classroom. After the Aviation Cadet graduated and obtained wings and a commission, he got about 60 hours additional flying in advanced trainers before flying a combat plane. In the early years of World War II a number of specialized twin engine aircraft were purchased for training bomber crewmen since almost all combat

32.CURTISS AT-5A A trainer version of the Curtiss P-1 Hawk series, the AT-5A shown in this company photo was one of 71 AT-4 and AT-5 airplanes using an engine of lower power than the fighters. The 35 AT-4s of 1927 used a Wright Hisso E liquid cooled engine of 180 horsepower, and the AT-5 and AT-5A series of 1927-8, 36 aircraft, employed the air cooled radial Wright Whirlwind of 220 horsepower. Shortly afterwards in 1929 the Army needed more pursuit planes and 52 of these trainers were re-engined with the more powerful 435 horsepower Curtiss D-12 engine and redesignated P-1D,E, and F Hawk fighters.

33.North American AT-6 An early version of the AT-6/T-6 Texan series with pre-World War II tail stripes in a company flight shot. The AT-6s were advanced trainer versions of the earlier BC-1 Basic Combat type, and the planes all US and many foreign pilots trained in prior to flying combat aircraft. Well over 10000 of these famous Texans were built; many were turned over to the US Navy as SNJ- advanced trainers and others went to foreign goverments. High speed was just over 200 mph using a Pratt and Whitney Wasp engine of 600 horsepower. T-6s also saw spotter service in Korea, and are used as latter-day racers and hobbyist airplanes.

bomber production was needed at the war fronts. Trainers for multi-engined pilots, navigators, bombardiers, and gunners were procured in great numbers because the overall combat aircrew training program was huge. As the war progressed into 1943 bomber production swelled, and combat types in early versions became available for training purposes. A total of over 550 Martin B-26Bs and B-26Cs became target towing aircraft as AT-23s and a large number of B-25 Mitchell bombers became AT-24s. Later the AT-24s became TB-25s post-war, thus ending the AT- series. After the war the replacement aircraft for the T-6 Texan was the North American T-28 Trojan of 1950, but this plane was considered as a basic trainer, not an advanced trainer. The jet-powered Lockheed T-33 became the USAF advanced trainer and later the T-38 Talon took over advanced training, these to be covered in another section.

34.BEECH AT-7 One of many military versions of the Beech 18 civil light transport, the AT-7 Navigator was produced in 1941 to 1943 as the Army plane for training student navigators. Produced in two major versions, AT-7 and AT-7C, with a slight engine change, a total of 1142 Navigator types were turned out, including five AT-7A floatplane or skiplane variants. A four ton gross weight aircraft, the AT-7 had a high speed of 220 mph, and provided positions for training three students at a time.

35.CURTISS AT-9 Unofficially called the Jeep, this Curtiss twin engined airplane provided future World War II pilots with multi-engine transitional training early in the war. A total of 791 of these all metal four place three ton gross weight machines were produced at the Curtiss Wright St. Louis, Missouri plant in 1941 and 1942 in two versions, AT-9 and AT-9A with minor differences. Powered by Lycoming radial engines of 295 horsepower each, the AT-9 had a maximum speed of just under 200 mph.

36.BEECH AT-11 An AT-11 Kansan bombardier trainer is shown in a company photograph. Another modification of the civil Beech 18, the Kansan had a bombardier station at the glassed-in nose and an internal bay for small practice bombs in the fuselage. Guns could also be fitted. A total of 1582 of these trainers were built in 1942 by Beech at Wichita, Kansas and were used until training versions of actual bombers became available later in World War II.

37. SEVERSKY 2PA-L The photo shows a 1937 Seversky 2PA-L export Convoy Fighter which later evolved, with another engine, into the Republic AT-12 Guardsman advanced trainer. The 2PA-L was a two seat variant of the Seversky P-35 Army pursuit airplane, but used a 1000 horsepower nine cylinder Wright Cyclone engine. It went to Russia, and later twenty more were sold to Japan in 1938. Sweden ordered 52 aircraft with the Pratt and Whitney Twin Wasp engine of 1050 horsepower, but in 1941 50 of these were requisitioned for the US Army as the AT-12 with a high speed of 285 mph and a gross weight of 7500 pounds. They were used mostly as utility hacks at various airfields.

37-A.CESSNA AT-17 A military advanced trainer version of the Cessna T-50 light twin engine commercial transport, the US Army AT-17 Bobcat was known in Canadian service as the Crane. After 33 similar AT-8 Bobcats were produced in 1941 mass production began in 1942 with a total of 1199 advanced trainer aircraft produced. Engines were Jacobs radials of 245 horsepower each. The 5200 pound gross weight aircraft had a top speed of 170 mph. Many of these planes became Army UC-78 light transports used in utility missions. Canada received over 500 for training purposes.

S38.TINSON AT-19 A militarized version of the popular pre-war Stinson Reliant, the AT-19 was built as an instrument trainer for the British Royal Navy. Five hundred Reliants powered by a Lycoming R-680 engine of 280 horsepower were produced for that service in 1942-3. Though the US Army AT-19 designation was applied for record purposes, none were used by the US military. The five place Reliant had a wing span of 41 feet 11 inches, a gross weight of about 4300 pounds, and had a high speed of 135 mph.

39.FAIRCHILD AT-21 A development of the earlier XAT-13 and XAT-14 crew and bombardier trainer prototypes respectively, the Fairchild AT-21 Gunner was introduced in 1942 as a specialized advanced trainer for bomber gunners with a dorsal power turret. Major production was intended by Fairchild, Bellanca, and McDonnell, but only 175 aircraft were manufactured, 106 of them by Fairchild since gunner training was later shifted to trainer versions of the actual bombers. Powered by two 520 horsepower Ranger engines, the six ton AT-21 had a maximum speed of 230 mph.

BASIC COMBAT

The Basic Combat (BC-) category of aircraft was short lived, used only from the late 1930s through 1940. Since the last aircraft in the Advanced Trainer (AT-) category was the Curtiss AT-5A of 1928, a converted single place pursuit, by 1937 almost ten years had passed without procurement of a new trainer having the general characteristics of a really modern combat aircraft, including operational equipment and reasonable performance. The decision was made to define a basic combat category covering a final training plane to indoctrinate a pilot trainee prior to combat squadron assignment. North American won the aircraft competition and produced their model NA-36 designated by the Army as a BC-1 in 1938, as it turned out the only BC- type ever put into production. Three BC-2s were delivered in 1938 as BC-1s with a modified powerplant including a three bladed propeller, and the single Vultee BC-3 prototype of 1940 was not honored with a production contract. The BC-1 aircraft design evolved into the AT-6 Texan advanced trainer; the AT- category continued through World War II, but the BC- category was not used again.

103.NORTH AMERICAN BC-1 A direct predecessor of the AT-6 advanced trainer, the BC-1 was the only aircraft produced in quantity in the Basic Combat category, this meant to duplicate characteristics of actual combat types. A total of 180 of these trainers were produced starting in 1938; a further 92 BC-1A versions came out in 1940, the last few of these being delivered as AT-6 types. Powered with a Pratt and Whitney Wasp engine of 600 horsepower, the 5200 pound BC-1 had a maximum speed of just over 200 mph. The BC-1 shown has a fabric covered fuselage; later AT-6s had all metal fuselages.

BOMBERS

The bombing airplane has always been considered the backbone of the Army Air Force and the USAF. Its ability to cause great destruction to enemy installations of all kinds made it the premier offensive aerial war machine. Shortly after the end of World War I the US Army adopted a scheme classifying bombers into Day Bombardment, Night Bombardment-Short Distance, and Night Bombardment-Long Distance. The only bombing airplanes developed in the early 1920s were in the second category of Night Bombardment-Short Distance. The two engined Glenn Martin bomber of 1921 was America's bombardment plane, carrying a ton of bombs for about 400 miles. About 130 of this type, made by various manufacturers, were purchased in the early 1920s. In 1921 the Martins made an air power point by sinking moored captured German warships off the east coast, proving General Billy Mitchell and his air enthusiasts' claim that airplanes could sink battleships, something to be shown again in World War II. By the late 1920s barely 60 bombers were left. In 1929 a considerable advancement in design was made represented by the Curtiss B-2 Condor, but only a dozen of the type could be procured.

The early 1930s were dominated by the Keystone bombers of which about 200 were built. Twin engined biplanes of six to seven tons gross weight cruising at 95 mph, their range was about 400 miles with a ton of bombs, or hardly a performance change from the earlier NBS- bombers. But they were relatively cheap and thus affordable in the minimal military budgets of the day.

In 1934 however, Martin came out with a new twin engined B-10 monoplane bomber which for its time was a world-beater incorporating all the latest technology of the day, and 135 of this type became the standard US Army bombing machine, making many record-breaking flights. The next year another revolution in bombers occurred with flight of the Boeing XB-17, the predecessor of the World War II Flying Fortress, and the US heavy four engined bomber was born with capability of carrying five tons of bombs for about 1400 miles. In the 1937 time period the US Army had to make a choice between buying a few four engine heavies or a lot more twin engined medium bombers represented by the Douglas B-18, a plane based on the commercial DC-3 and able to carry two tons of bombs for 1200 miles. The Army bought only 13 Boeings and purchased 350 Douglases. By 1939 a few more B-17s were procured, and in 1940 the Army bombers were classified as light, medium, and heavy, as represented respectively by the Douglas A-20 Havoc, the Douglas B-18, and the Boeing B-17. The B-18 was quickly being replaced in 1941 by the B-25 and B-26 mediums, and the modernized B-17 was supplemented by the new long range B-24 from Consolidated. These were the bombers of World War II, turned out by the hundreds with the B-24 noteable as the most-produced US warplane of all time with almost 18500 turned out. The story of USAAF wartime bombers is well known. Defensive armament increased steadily as did weights with speed decreasing. The theory of successful high altitude pin point day bombing with tolerable losses by unescorted heavy bombers espoused early by US planners was shown to be wanting. Escort fighters were needed in great quantity. The sum totals for all major US bomber types used in European combat shows they dropped almost 1.4 million tons of bombs and shot down almost 10000 enemy planes while losing the same number of their own. Overall bombing effectiveness results are somewhat controversial, but there is no doubt the destruction was tremendous. The same aircraft types continuously bombed Japanese Pacific bases, however Japan's fate was sealed with the advent of the very heavy long range bomber, the Boeing B-29, with about 2000 produced by a major industrial consortium, and capable of carrying 10 tons of bombs for over 3000 miles. Most significantly the B-29 could carry the atomic bomb and ended the war by dropping two upon Japan. Post-war the re-engined B-29 became the B-50.

The final piston engined bomber, one with intercontinental performance in nuclear bomb delivery, was the Convair B-36. A 200 ton behemoth with a 400 mph high speed and 40000 foot ceiling, the B-36 had a combat radius of about 4000 miles with five tons of bombs. The B-36 was a US Cold War threat for several years and served as an effective deterrant. Jet engines took over in bomber design first in 1947 with a straight wing four jet North American B-45 Tornado, and then with the radical swept wing six jet Boeing B-47 Stratojet of the 1950s, a 600 mph aircraft capable of carrying five tons of bombs for a 2000 mile radius with 2000 built. Aerial refueling capability was installed to increase combat radius significantly. Medium jet bombers of the 1950s and 1960s included the B-57 and B-66, both in many versions.

The close follow-on to the B-47 was the eight jet Boeing B-52 Stratofortress very heavy bomber with, in 1956, a demonstrated capability of dropping a thermonuclear bomb and the capability, without aerial refueling, of carrying five tons of bombs for a combat radius of about 4000 miles and a much heavier load for shorter distances. Strategic Air Command had over 600 B-52s in the early 1960s; a decade later after their use as conventional bombers in the Vietnam War the number was nearer 450 aircraft. A few remained for limited conventional use in the Gulf War.

The very high speed high altitude Convair B-58 delta wing Hustler bomber of the early 1960s was procured in quite limited numbers since USAF dollars were being spent heavily upon intercontinental ballisic missiles, and high altitude bombers were becoming increasingly vulnerable to missile attack from the ground

40.CURTISS JN-6HB A version of the famous Curtiss Jenny, this one a Wright Hispano engine powered day bomber trainer. The same size as the earlier Curtiss OX- powered JN-4 Jenny, the JN-6 was heavier and much more powerful (150 horsepower versus 90) than the JN-4 airplanes. A total of 154 of these bomber trainers were produced by Curtiss starting in the summer of 1918. Note the racks for small bombs under the fuselage.

and the air. Replacing the B-52 has been difficult. The Mach 3.0 B-70 bomber program was aborted in the mid-1960s; the FB-111A of 1970-71 was produced only in small numbers, and the B-1A of the 1970s was cancelled in favor of missile development, only to be revived in modified and updated form years later. But less than 100 B-1B Lancers were produced, and did not participate in the Gulf War. The Northrop Grumman B-2 "Stealth Bomber" of very low radar cross-section and unusual configuration costs so much that only 21 aircraft are in the program, and questions are continuously raised as to what they are to be used for. It is thus unclear what the future of the USAF manned bomber may be.

41.MARTIN MB-1 One of ten Martin MB-1 (also known as GMB for Glenn Martin Bomber) US Army day bombers that first flew in the summer of 1918. A 71 foot 5 inch wingspan three place biplane using two 400 horsepower Liberty 12 water cooled engines, the Martin grossed out at 10225 pounds. Nose, dorsal, and ventral gun positions were provided, and a thousand pounds of bombs could be carried. High speed was about 115 mph. Many conversions for special testing were made on this batch of Martin bombers.

42.MARTIN MB-2 A larger and heavier development of the MB-1, the MB-2 was later redesignated as an NBS-1, standing for Night Bomber-Short Distance under the then-current designation system. The initial order was for 20 MB-2s from Martin. These aircraft weighted six tons gross, used two 420 horsepower Liberty engines, had a crew of four, and carried twin .30 caliber machine guns at nose and dorsal positions and a single ventral .30 caliber weapon. Increased bomb loads could be carried. Additional NBS-1 bombers were produced by Martin (15), Curtiss (50), and LWF (35).

43.CURTISS B-2 With the experience of building 50 NBS-1 bombers Curtiss embarked on a new larger bomber design in 1926 and two years later delivered 12 additional B-2 types to the Army with technical improvements. With a 90 foot span and a gross weight of 16500 pounds the B-2 was good sized for its day. The two 600 horsepower Curtiss Conqueror engines provided a high speed of 133 mph. A distinguishing feature was location of a gunner position at the rear of each engine nacelle; another was the biplane horizontal tail.

44.KEYSTONE B-3A In the late 1920s and early 1930s the Keystone Aircraft Company in Bristol, Pennsylvania was the bomber builder with a long series of six to seven ton gross weight twin radial engined biplanes. One of these was the B-3A type shown, one of 36 five place bombers of 1930 originally designated an LB-10A light bomber. Powered with Pratt and Whitney Hornet engines each of 525 horsepower, the B-3As had a maximum speed of 114 mph and cruised at 98 mph.

45.KEYSTONE B-6A Another of the Keystone bombers, and the final version, the B-6A biplane was ordered to the extent of a dozen aircraft in 1932 after five service test Y1B-6 bombers proved themselves the previous year. These aircraft weighed about 13300 pounds loaded and had more power, using two new 575 horsepower Wright Cyclone radial engines which upped high speed to 120 mph. The Keystone bomber fleet was first line Army equipment until the new Martin metal monoplane bombers got into service.

46.DOUGLAS XB-7 This Army photo shows the experimental Douglas XB-7 twin engine four place high gull wing five ton gross weight aircraft which the Army could not decide was a light bomber or an observation plane, originally labeling it an XO-36. An additional 13 planes of this type were ordered for service test, some as Y1B-7 bombers and others XO-35 or Y1O-35 observation types. Never reaching full production status, this 1931 to 1933 airplane pioneered some modern features such as metal construction and retractable landing gear. Two 600 horsepower Curtiss Conqueror liquid cooled engines provided a high speed of about 180 mph.

47.BOEING YB-9 A real step forward in bomber design, with all metal construction and a cantilever monoplane wing, the twin engined YB-9 and Y1B-9 were Boeing company-financed ventures. Powered by 600 horsepower Pratt and Whitney Hornet B radial engines, the first flight of this bomber type took place in the spring of 1931. Liquid cooled Curtiss Conqueror engines were also tested on the aircraft. Though the four place B-9 type could outpace contemporary fighters at over 180 mph only five service test bombers were produced for the Army.

48.MARTIN YB-10 The Martin YB-10 bomber shown in the photo was one of 14 service test aircraft ordered in 1933 after successful testing of a single much revised prototype built as a private endeavor by the Martin Company. Revolutionary for the time with cantilever all metal construction , crew enclosures, internal bomb bay, and retractable landing gear, the Martin had single .30 caliber guns in a nose turret and dorsal and ventral positions. The B-10 became the Army's standard bomber in the mid-1930s, a total of 104 YB-10A and B-10Bs were ordered. The high speed of over 210 mph could be achieved by this new machine using two 750 horsepower Wright Cyclone engines along with a 190 mph cruise speed.

49.MARTIN YB-12A Seaplane The Martin B-12 series was similar to the B-10 bomber but powered with two 775 horsepower Pratt and Whitney Hornet engines. A total of 32 B-12 types were produced in 1933 and 1934 and, like the B-10s, were standard Army bombers. Performance was generally similar also. The B-12A shown on floats was part of a US coastal defense force after it was agreed in early 1931 that the Army would take over that mission from the Navy. The wheels on the floats were for aircraft handling purposes.

50.BOEING XB-15 The single XB-15 long range bomber was an experiment in building a very large aircraft with a bombing range of up to 5000 miles. It was ordered in mid-1935 and first flown a little over two years later. With a wingspan of 149 feet and gross weight of over 70000 pounds, the XB-15 was a behemoth for its day, but suffered from being underpowered by four Pratt and Whitney Twin Wasp engines of only 1000 horsepower each. High speed was just under 200 mph. In World War II the XB-15 carried Army cargo as the XC-105.

51.BOEING 299 The rollout of the prototype Boeing 299 Flying Fortress in July of 1935 was a significant event in US bomber history. Contrast the sleek lines of the new 299 with those of the parked automobiles. Later called the XB-17, the four engine eight place Fortress spanned almost 104 feet, weighed 43000 pounds loaded, and was powered by Pratt and Whitney Hornet engines of 750 horsepower each, providing a maximum speed of 250 mph. There were no engine turbosuperchargers. The aircraft crashed in October, 1935 when a takeoff was mistakenly attempted with locked controls.

52.BOEING Y1B-17 A striking view of a Y1B-17, one of 13 service test bombers ordered in early 1936, a big change being use of four Wright Cyclone engines. Other changes included addition of another crew member and a slight increase in weight and performance. All Y1B-17s except one aircraft served with the 2nd Bombardment Group at Langley Field, Virginia starting in 1937 and made many well-publicized flights. The Y1B-17 was the last Fortress variant without turbosupercharged engines.

53.BOEING B-17C After experimentation with a single Y1B-17A and production of 39 B-17B models Boeing put out 38 B-17C Fortresses in late 1940 similar to the B except with flush side gun positions and addition of a "bathtub" lower gunner station. Fitted with turbosupercharged Cyclones of 1200 horsepower each, the C model could attain over 300 mph at altitude. Another change was use of .50 caliber guns instead of .30 caliber at all positions.

54.BOEING B-17F This beautiful Boeing flight shot of a new ten place B-17F Flying Fortress illustrates some of the features added in the developed airplane including the dorsal fin, ribless Plexiglas nose, and forward upper power turret. When 512 B-17Es had been made mass production started on this F model after a first flight in the spring of 1942. Boeing produced 2300, Douglas 600, and Vega 500 of this 55000 pound to 65000 pound gross weight bomber. Extra fuel tanks and external bonb racks were added on late F versions so a maximum bomb load of over ten tons could then be carried.

55.BOEING B-17F Another company photo of a new B-17F Fortress illustrates belly ball power turret, the semi-retractable main landing gear and the location of engine turbosuperchargers in the under portions of a nacelle. The early F did not carry wing bomb racks. Note the photo is early enough that no bars are carried in the national insignia. Power was provided by four 1200 horsepower Wright nine cylinder Cyclone engines.

56.BOEING B-17G The final production version of the Boeing Flying Fortress, the B-17G, is shown running up in this company photo. The main addition in the G was a chin turret forward to fix a weakness of the B-17 against head-on attacks from fighters. The production team of Boeing, Douglas, and Vega produced a total of 8680 G models starting in 1943, making it the most-produced version. Some Gs were converted to B-17H air-sea-rescue versions carrying a lifeboat under the fuselage.

57.DOUGLAS B-18 A bomber version of the Douglas DC-2 civil transport, the stubby-nosed B-18 was ordered from Douglas in 1936 to replace Martin B-10 and B-12 bombers. The Army had decided to buy a large number (133) of the cheaper two engine bombers rather than procure a few more expensive B-17 Fortresses. The six place B-18 was powered by two 930 horsepower Wright Cyclones but was slow with a high speed of only 217 mph. It was also lightly armed with three .30 caliber machine guns, one each in nose, dorsal, and ventral positions.

58.DOUGLAS B-18A A revised version of the B-18 with engines of 1000 horsepower and increased gross weight along with a change in nose arrangement putting the gunner below the bombardier, but no real change in performance. A total of 217 were built in 1938 and the aircraft became the standard Army bomber, serving right up until Pearl Harbor and afterwards. The photo shows a B-18A in 1941 warpaint. The B-18s saw no real offensive bomber action; some were destroyed at Pearl harbor by the Japanese; others went on maritime patrol duty.

59.DOUGLAS B-18B A post-war photo of a weary B-18B Bolo, one of 122 converted from B-18As for wartime coastal patrol against marauding German submarines. Early ASW gear was installed on these airplanes; they served as stalwarts in the early months of World War II until replaced by more modern types.

60.DOUGLAS XB-19A A product of Army interest in very large long range bombing airplanes, the Douglas B-19 "Hemisphere Defender" was a single airplane with a wingspan of 212 feet and a gross weight of 80 tons. Used during wartime as a transport after a 1941 first flight, the XB-19 was purely an experiment. Originally under-powered by four 2000 horsepower Wright R-3350 18 cylinder radial engines, it was later revised as shown in the photo for 2600 horsepower Allison V-3420 24 cylinder liquid cooled inline W-type powerplants.

61.CONSOLIDATED B-24D The first mass production model of the big Liberator bomber, the one pictured a postwar rehabilitated aircraft. The deep capacious fuselage enclosed a bomb bay for up to six tons of bombs; other features included a new Davis wing with a considerably higher aspect ratio than the B-17 for long range and slab-like twin vertical tails. Four turbosupercharged 1200 horsepower Pratt and Whitney Twin Wasp engines powered the B-24D giving a high speed of 300 mph. The first B-24D came off the line in January of 1942. A total of 2738 D models were produced, a few by Douglas.

62.CONSOLIDATED B-24D Lines of the B-24D Liberator are well shown in this flight photo. Armament comprised two twin .50 caliber power turrets, one atop the fuselage forward and the other at the tail. Manually served .50s were mounted at two waist, nose, and ventral positions, though armament could vary. B-24D models first saw action in mid-1942. Export models went to England as LB-30s which were used in both transport and coastal and deep water patrol missions. Long B-24 range narrowed the areas where German U-boats could operate free from air attack.

63.CONSOLIDATED B-24J After over 4300 B-24E, G, and H Liberators had been put out by a combination of Consolidated, North American, Douglas, and of Ford at the huge new Willow Run plant, the most produced version was to be the B-24J with fully 6678 airplanes fabricated by the four manufacturers starting in mid-1943. This 12 place 30 ton gross weight bomber had a high speed of 290 mph. The J model shown in the photo had a nose and lower ball power turret added to those at upper fuselage and tail positions along with manually operated guns at the fuselage waist positions.

64.CONSOLIDATED B-24M The final Liberator production type, a B-24M, came after 1657 B-24L aircraft had been turned out by the consortium. Armament changes characterized the B-24M, specifically a light weight tail turret. Strangely, perhaps, more B-24s were manufactured than any other US warplane. One might think off-hand more of a single engine fighter type would have been turned out rather than a 30 ton four engine bomber. The B-24 had a reputation as a more difficult aircraft to fly and as more vulnerable to gunfire than the B-17.

65.NORTH AMERICAN B-25 A redesign of the NA-40 and NA-40B attack bomber types entered by North American in a 1938-9 Army competition, the B-25 Billy Mitchell medium bomber was a five place 27300 pound gross weight mid-wing aircraft powered by two Wright R-2600 engines each of 1700 horsepower. High speed was over 320 mph. Twenty-four B-25 and 40 B-25A airplanes were produced in 1940-41 by North American, the latter incorporating crew armor and self sealing fuel tanks. The aircraft had no power turrets, but included a tail gun position.

66.NORTH AMERICAN B-25B The photo shows a B-25D variant modified back to the B-25B configuration to be used as a USAF Museum exhibit. Most all of the 119 B-25Bs had been delivered by the end of 1941, these distinguished by an upper and lower twin .50 caliber gun turret, the latter retractable, with waist and tail gun positions eliminated. The bomb bay could accomodate up to 2400 pounds. Heavier and slower than the earlier models, the B-25B was chiefly distinguished by use of 16 of these aircraft for the Doolitle raid on Japan starting from the aircraft carrier Hornet in early 1942.

67.NORTH AMERICAN B-25J After a total of over 5300 B-25C, D, G, and H Mitchell models had been produced North American in 1943 swung into manufacture of the B-25J with a crew increase to six and gross weight upped to 35000 pounds dropping high speed to about 275 mph. Over 4300 J variants alone were produced. Heavily armed with 12 or more .50 caliber machine guns, the J had two in the transparent nose, four package guns fixed alongside, two in an upper turret, two waist, and two in the tail. A bomb load of up to 3200 pounds could be carried internally.

68.NORTH AMERICAN B-25J A post-war civil B-25J rehabilitated in Army colors runs up ready to perform at a 1983 airshow. Just under 10000 Mitchells were manufactured. Many were provided to the air forces of other countries under Lend Lease including Britain, Russia, and China. Some B-25Js were supplied to the US Navy as PBJ-1J patrol bombers.

69.NORTH AMERICAN B-25J Another post-war rehabilitated B-25J Mitchell at a 1988 airshow illustrates the attack variant with a solid metal nose. Terrific forward firepower could be generated by multiple fixed .50 caliber machine guns in this nose, often as many as eight, and with fuselage package guns on the side, sometimes 12 fixed guns firing forward. Earlier versions like the B-25G and H carried a 75 mm cannon fixed in the nose in addition to four machine guns.

70.MARTIN B-26A From one of the two early production batches of Martin Marauders, this B-26A was one of those used in early South Pacific action starting in April, 1942. After 201 Marauders had been produced in a design right off the drawing board, Martin followed with 139 B-26As. This graceful twin engine seven place medium bomber using two 1850 horsepower Pratt and Whitney Double Wasp engines had achieved a high speed of 310 mph, but with a small wing of 65 foot span had a high wing loading and the resultant high landing speed for its time. It had a bad reputation among many pilots.

71. MAR TIN B-26B/AT-23A The B-26B Marauder was the first model put into real mass production in the spring of 1942 with Martin manufacturing 1883 aircraft. After 641 had been built wingspan was increased to 71 feet which added 50 square feet to wing area. Gross weight increased to 17 tons and Double Wasp engine power was increased to 2000 horsepower. Most importantly armament was revised to include up to eight or 12 .50 caliber guns. The aircraft shown is an AT-23A advanced trainer version , one of 208 B-26Bs converted for target towing and bomber crew advanced training.

72.MARTIN B-26C (civil) A post-war civil conversion of a B-26C Marauder. The B-26C, 1235 of which were built in a second-source plant at Omaha, Nebraska, was similar to the B version. The clean lines of the Marauder are evident. The B-26Cs had the larger wing with slotted flaps like the later Bs, and this change was needed because gross weight was up to 19 tons. A few B-26Cs were converted to AT-23B advanced trainers.

73.MARTIN B-26F An excellent view of the B-26F Marauder, 300 of which were built by Martin at Baltimore, Maryland. This model had two significant changes-four fixed forward-firing fuselage .50 caliber guns were provided as standard (though not shown on the aircraft in the photo), and the wing nose-up incidence angle was increased significantly, and enough to be noticeable, this again to lower takeoff and landing speeds. Maximum speed was down to about 275 mph. B-26 Marauders provided excellent service with the USAAF Eighth and Ninth Air Forces in the European Theatre.

74.MARTIN Marauder III A company photo of a Marauder III for Britain, this version similar to the USAAF B-26F and G, running up at the factory. Twin .50 caliber dorsal turret and twin gun tail positions are readily seen, along with a lower waist gun port. A total of 350 Marauder IIIs were sent to the British. Marauder bombers flew some very tough missions; over 900 USAAF aircraft of the B-26 type were lost in combat operations. The last Marauder version was the B-26G with minor changes; 893 Marauders were produced in 1944.

75.BOEING B-29 The B-29 Superfortress, an early camouflaged model of which is shown in this USAAF flight photo, was indeed the super bomber of World War II, and 4221 of these four engined heavy bombers were produced by a consortium of Boeing in Renton, Washington and Wichita, Kansas, Bell in Atlanta, Georgia, and Martin in Omaha, Nebraska. Used only against Japan, and now known mainly for being the delivery vehicle for two atomic bombs dropped on that country, the B-29 weighed 60 to 70 tons, spanned 141 feet, and had 8800 installed horsepower in four Wright R-3350 18 cylinder engines each having two superchargers, and employed a crew of eleven. Up to ten tons of bombs could be carried.

76.BOEING B-29 The B-29 Superfortress Pacasan Dreamboat was used for a special record-breaking Pacific long range flight just after the war. The size of the bomber may be judged from the size of the men gathered alongside. Dual wheels were needed on each leg of the tricycle landing gear to support the high gross weight of the aircraft. The ten gun defensive armament was concentrated in four remote controlled power turrets on the fuselage and a single directly controlled tail turret which originally was equipped with two .50 caliber machine guns and a 20mm cannon. The fuselage was divided into three pressurized compartments for the crew.

77.BOEING RB-29 One of the last batches of B-29s built at Boeing, Wichita, this aircraft was used for reconnaissance missions. Bomb bay doors are open; notice the split bay arrangement to allow the mid-wing structure to carry through the fuselage. Note the four blade propellers employed as opposed to the three blade models in the previous photo. Range of the bomber version was over 3100 miles with ten tons of bombs; range of reconnaissance versions could be over 5000 miles.

78.BOEING YKB-29J One of two RB-29J reconnaissance Superfortresses converted to flying tankers for aerial refueling. Earlier B-29 aircraft had been revised to provide probe and drogue type refueling KB-29M types, but the YKB-29J aircraft tested the new Boeing-developed flying boom refueling method. This equipment can be seen installed at the tail of the aircraft. The arrangement proved successful, and 116 aircraft were modified to KB-29P aerial tanker configuration with flying booms starting in 1949. The last operational B-29 types were finally retired in 1960.

C79.ONVAIR B-32 The B-32 Dominator was built as a hedge against failure of the Boeing B-29 in the World War II heavy bomber development program. After testing a couple of twin tailed pressurized XB-32s along with a single tailed third prototype , Convair started delivery of 115 production Dominators in late 1944. The production aircraft were unpressurized and had ten .50 caliber machine guns in power turrets. B-32 bombers got in barely a mission or two before the end of the war; they were 12 place 112000 pound gross weight high wing aircraft with a maximum speed of just over 350 mph, and were powered with four Wright R-3350 engines each of 2200 horsepower.

80.CONVAIR B-36A A contest at the Convair plant resulted in the unofficial name "Peacemaker" for this huge aircraft, but the name was never officially approved or used. This bomber resulted from a USAAF design competition of 1941 for an aircraft able to carry a five ton bomb load for a 5000 mile radius of action. The XB-36 was rolled out just at the end of World War II, and first flew a year later in 1946. Another year passed before a production B-36A, the first of 22, appeared. Powered by six 3250 horsepower Pratt and Whitney Wasp Major engines in a pusher propeller arrangement, and with a gross weight of 155 tons and wingspan of 230 feet, the B-36A could not quite make a high speed of 350 mph.

81.CONVAIR B-36B The picture shows a B-36B airplane, one of 73 produced and the first true operational version, after updating with four jet engines to B-36D configuration. The B-36B had heavy defensive armament with a total of eight turrets each with two 20mm cannon, two of the turrets retractable and controlled remotely. Most all (64) of the 73 Bs were converted to B-36Ds. The US Navy was unsuccessful in trying to sink the B-36 program in 1949, and production of the huge bomber continued. The Strategic Air Command now had an intercontinental delivery vehicle for atomic weapons.

82.CONVAIR B-36J The final B-36 version, of which 33 were built with increased fuel capacity, bringing the total B-36 production to over 380 airplanes, some of which were bomber and others reconnaissance aircraft. The J model, first flying in 1953, had six 3800 horsepower Wasp Major engines plus four General Electric J-47 jet engines each of 5200 pounds thrust. Gross weight was up to 205 tons, though some aircraft were stripped down to operate at extra high altitude. The B-36J had a high speed of just over 410 mph at 37000 feet altitude and could operate at about 40000 feet. It could carry five tons of bombs for a radius of about 4000 miles. The B-36 served until the mid- to late 1950s.

L83.OCKHEED B-37 A follow-on to the Lockheed Hudson, the USAAF B-37 and B-34 twin engine bombers along with the export R-37 Ventura were military derivatives of the civil Lockheed Model 18 Lodestar transport. The five place B-37 shown, of which 18 were delivered in 1942, used Wright R-2600 engines each of 1700 horsepower, and had a high speed of just under 300 mph. The B-34 and R-37 used by the British grossed at about 14 tons and were powered by Pratt and Whitney Double Wasps each of 2000 horsepower, giving a high speed of 315 mph. Several hundred of the Double Wasp powered bombers were used by the USAAF and the RAF. Bomb loads of up to 3000 pounds could be carried. Armament consisted of two fixed nose guns, two in a dorsal power turret, and one in a ventral position. The US Navy version was the PV-1.

84.NORTH AMERICAN B-45A The first US production jet bomber, the four engine straight wing North American four place B-45A Tornado was flown as a prototype in early 1947. Looking like a twin jet aircraft because two engines were contained in each nacelle on the wings, the later As used four General Electric J-47 jets of 5200 pound thrust each and had a high speed low down of 570 mph and 500 mph at high altitude. Since it was felt a fighter attack could come only from behind, the sole defensive armament was two tail .50 caliber guns. Normal gross weight of the aircraft was over 40 tons. Ninety-six B-45As served from 1948 into the later 1950s.

85.NORTH AMERICAN B-45A Ground view of a B-45A Tornado bomber shows details of the nacelles each housing two jet engines, the tricycle landing gear with main wheels folding for retraction into inboard wing sections, and the clean lines of the fuselage interrupted only by the bubble canopy for pilot and copilot. The B-45A could carry five tons of bombs for a radius of over 500 miles. High fuel consumption of early jet engines penalized bomber range performance.

86.NORTH AMERICAN RB-45C The 1949-50 reconnaissance version of the Tornado with multiple cameras installed, 33 of which were ordered as a natural development of a very fast bomber for its time. RB-45C aircraft were used over Korea. Longer ranges were obtained by carrying extra fuel in the bomb bay and in very large wingtip fuel tanks as shown in the photo. Water injection was added to the jet engines to enhance takeoff performance. Gross weights up to 55 tons were employed in these reconnaissance airplanes, of which over 26 tons could be jet fuel.

87.BOEING XB-47 After over three years of design studies including examination of German research on high speed flight, Boeing in December, 1947 first flew a revolutionary new US bomber, the XB-47. The Stratojet was a three place six engine aircraft of 125000 pounds gross weight with the new design features of a 35 degree swept back wing and a bicycle and outrigger landing gear. The General Electric jet engines each of 3750 to 5000 pounds thrust were podded under the high wing, two together inboard and one far out on the wing on each side. The high speed of the XB-47 was 580 mph at medium altitude with the initial engine installation and over 600 mph with later engines.

88.BOEING B-47A A beautiful Boeing flight shot of a B-47A, ten of which were ordered as service test aircraft ahead of production models and which appeared in mid-1950. These 75 ton gross weight aircraft had assistance in takeoff from 18 solid fuel rockets mounted along the sides of the aft fuselage and landing assistance from a large drag parachute extending from the tail. Only tail armament was included. When the aircraft took off static droop of the wings changed to dihedral as the aircraft weight shifted to the wings.

89.BOEING B-47E The B-47E Stratojet of 1953 was the major production version of the swept-wing Boeing bomber after 398 of the initial production B-47Bs had been turned out. Lockheed and Douglas assisted Boeing in producing a total of over 1580 B-47E bomber and RB-47E reconnaissance variants. With the six General Electric J-47 engines upgraded to 6000 pounds of thrust each and gross weight of about 100 tons the 600 mph B-47E was a formidable threat with ability to carry a nuclear weapon over long ranges, particularly when assisted by tanker airplanes using the Boeing flying boom aerial refueling system.

90.BOEING TB-50A The B-50 Superfortress was a direct post World War II development of the B-29. There were many changes, but those most apparent externally were installation of new Pratt and Whitney four row 28 cylinder Wasp Major engines of 3500 horsepower each in place of the B-29's 2200 horsepower Wrights, and a larger vertical tail incorporating rudder boost. A total of 79 B-50As were delivered to Strategic Air Command in 1947 and 45 improved B-50Bs followed in 1949-50. The B-50A had a gross weight of 84 tons and a maximum speed of 385 mph. It was armed with 12 .50 caliber and one 20mm guns, and could carry up to 10 tons of bombs. The TB-50A training version shown was utilized for training bomber crews.

91.BOEING B-50D This version of the B-50 Superfortress was the most produced, 222 being delivered by Boeing after a mid-1949 first flight. The most important changes were in the fuel system including large internal tanks added and a single point refueling receptacle up front compatible with the flying boom system. In the photo a B-50D bomber is being refueled by a KB-29P tanker aircraft so that Superfort combat radius could be greatly extended. B-50D performance was otherwise similar to that of earlier models.

92.BOEING KB-50K As the B-50 design became obsolescent for the front line bomber role the Superforts were themselves converted into aerial tankers to be used for refueling aircraft of Tactical Air Command. Many of these aircraft carried jet engine pods underwing, one on each side, to boost speed capability to match TAC fighter minimum refueling speed. Probe and drogue refueling systems were incorporated. The KB-50 aircraft were employed for this purpose in the late 1950s and early 1960s.

93.BOEING B-52 The Boeing Stratofortress, shown taking off in the photo, was in appearance a direct larger development of the B-47 intended to carry a hydrogen bomb over intercontinental ranges. Powered by eight Pratt and Whitney J57 turbojets in early models, the 185 foot span aircraft first flew as a YB-52 in April of 1952. Three B-52As of 1954 initiated production with 1000 gallon drop tanks underwing along with provision for aerial refueling. The B-52B, of which 50 were built, could be used for bombing or reconnaissance, as could later B-52C aircraft with higher gross weight (200 tons) and more fuel. In late 1955 production started on 89 Wichita-built B-52Ds, 58 B-52Es, and later 45 B-52Fs. In the 1950s Seattle built 81 B-52Ds, 42 B-52Es, and 42 B-52Fs.

94.BOEING B-52G The first B-52G Stratofortress is shown in flight with a T-33 chase plane. The G model incorporated major revisions including a new vertical tail and fuel system changes, and flew first in late 1958. It had a greater range than previous B-52 aircraft. The final Stratofortress version was the B-52H with a change to eight Pratt and Whitney TF-33 turbofans each of 17000 pounds thrust. Gross weight was 488000 pounds and high speed was just over 650 mph . A 20mm radar-directed Vulcan cannon was installed in the tail. B-52 aircraft were used quite extensively for conventional bombing in the Vietnam War. A total of 744 B-52s were produced from 1952 to 1962, and some served into the 1980s.

95.MARTIN B-57B The Martin B-57 was unusual in being a basic British design adapted for production in the US by Martin, the first of 67 US-built B-57 light tactical bombers flying in mid-1953 using two Wright J-65 jet engines. The photo shows the next production model, the B-57B of 1954 with the crew of two in tandem under a bubble canopy and a rotary bomb bay door. The B spanned 64 feet, grossed out at 55000 pounds, and had a high speed of 580 mph. The Canberra carried up to three tons of bombs and employed a fixed forward firing armament of multiple .50 caliber or 20mm weapons. The aircraft served Tactical Air Command into the 1960s.

96.MARTIN B-57B A company flight view of one of 202 B-57B Canberras shows the broad wing with tip fuel tanks along with underwing store stations to carry bombs or rockets. Engines were Wright J65s of 7200 pounds thrust each, these again an American conversion by Wright Aeronautical of the British Armstrong Siddley Sapphire jet engine. A dual control version of the aircraft was the B-57C with 38 built, and later 68 B-57E target tug variants were produced. The Canberra saw some Vietnam War service.

97.MARTIN RB-57D A special high altitude version of the Canberra, of which 20 were manufactured, is shown in the photo. The big change was a new wing of over 40 feet greater span to increase both range and altitude capability for special snooper missions over Russia and China. Cameras and special electronic sensing gear items were carried in these reconnaissance Canberras.

98.CONVAIR B-58A The three place Hustler bomber was a unique delta wing design with four podded afterburner-equipped General Electric J79 jet engines each of 10300 pounds thrust, a gross weight of 82 tons, and a Mach 2 top speed at very high altitude. First flown in late 1956, 116 of the Hustlers were produced from the late 1950s to the early 1960s. A unique feature of the B-58A was the external releasable fuel and nuclear weapon pod located beneath the fuselage. Eight of the aircraft were put out as TB-58A dual control trainers.

99.CONVAIR B-58A A ground view of the B-58A shows the under-fuselage pod and its fairing and the long landing gear needed for pod clearance. The pilot sat up front, the bombardier-navigator just behind, and then the defense systems operator, the latter controlling a 20mm Vulcan high rate of fire tail cannon and the electronic defense gear. The B-58s operated through the 1960s in two bomber wings, but went out of the active inventory by the end of that decade.

100.DOUGLAS RB-66B The three place Douglas B-66 Destroyer tactical bomber was the USAF counterpart to the Navy A-3 Skywarrier with the naval carrier compatibility equipment omitted and Air Force equipment added. The twin jet swept wing 42 ton bomber had a high speed of 630 mph and could carry five tons of bombs over a radius of about 1000 miles. Coming out in 1954, versions of the B-66 were employed throughout the Vietnam War mostly as radar countermeasures aircraft. A total of 294 B-66 and RB-66 reconnaissance models in various versions were produced.

101.ROCKWELL B-1B After construction of four B-1A prototypes of a four place four engine 195 ton 137 foot span Mach 2 swing-wing strategic bomber, with a first flight in October,1984, production deliveries of a considerably modified B-1B Lancer version started in July, 1985 to the 96th Bomb Wing at Dyess AFB in Texas. A long range multi-role strategic bomber, the B-1B has a maximum range of 7450 miles and a maximum conventional weapon load of 64000 pounds. Cost of a B-1B bomber has been estimated at $280 million. There are currently 97 B-1B Lancer bombers in the USAF inventory. None participated in the recent Gulf War, presumably because of engine problems at the time; B-52s substituted for them.

NORTHROP B-2 The radical Northrop (now Northrop-Grumman) B-2 bomber shown in the photo is the latest design aircraft intended to drop nuclear bombs on an enemy. It employs a shape and construction materials to provide a minimal radar cross-section so as to avoid detection. Only 20 B-2 aircraft have been built, and since these few have cost an estimated $44 billion, making them the most expensive aircraft ever built it is likely there will be no more. In 1995 the Pentagon concluded that shrinking defense budgets made buying more B-2s impractical, and in February, 1996 President Clinton decided against building more. The $493 million authorized by Congress in the 1995 defense budget was to have been spent upgrading existing aircraft. Recently the president decided to order the Air Force to increase the B-2 fleet to 21 by upgrading one older test model and spend the $493 million for this work. The B-2 was first flown in July, 1989. It has a wingspan of 172 feet, a range of 6900 miles, can carry up to a 50000 pound weapon load, and has a cruising speed of about 550 mph. Normal takeoff gross weight is 371330 pounds with a maximum of 400000 pounds. With a normal crew of two (three optional) the B-2 is equipped with four 19000 pound thrust General Electric F118 non-afterburning turbofan engines. Two side by side weapons bays have a total capacity of 16 SRAM II missiles, 16 free-fall B83 nuclear bombs, or 80 1000 pound conventional bombs.

BASIC TRAINER

Basic trainers were produced for the U.S. Army from 1930 through the early part of World War II. These aircraft were employed for training prospective Army pilots in an intermediate stage between primary and advanced flight training. In the early 1930s Douglas two place observation biplanes cornered the basic trainer market with O-2 and O-32 conversions to BT-1 and BT-2 models. In 1936 two place low wing monoplanes with fixed landing gear and enclosed cockpits were purchased in the form of North American BT-9s and a few Seversky BT-8s. As World War II started in Europe, Vultee entered the market and walked away with basic trainer production for the huge pilot training program starting up with their BT-13 and BT-15 types. In 1941, just before the U.S. entered the war, five Army airfields were doing basic training. Because of the very tense situation at that time, basic training had been cut from 16 to 10 weeks and average flying time from 105 to 70 hours. Washouts were only about 14 percent. The first 20 hours was about half dual and half solo in the trainee's new 450 horsepower low wing basic trainer. There was strong emphasis on precision flying and later on advanced maneuvers, and finally night and instrument flying. The cadet who completed basic training was about 99 percent certain to earn his commission and wings.

After World War II it was decided to try and design a basic trainer which would serve both primary and basic training functions with the North American Trojan of 1949 being selected initially as an XBT-28, later revised to just T-28 when the new system came into effect. It was later found that this aircraft was satisfactory as a basic trainer, but too much airplane for a primary trainer. In any case there were no more basic trainers labeled in the BT-system.

104.DOUGLAS BT-2C A 1930-32 basic trainer variant of a long line of two place Douglas Army observation airplanes, the BT-2C was one of 20 improved models with a new landing gear after 146 BT-2B, and 31 BT-2 and BT-2A trainers had been produced, the latter being trainer versions of O-32 observation aircraft. Some BT-2Bs were converted to instrument trainers, while a few others were employed in 1940 as target drones. The BT-2 had a 450 horsepower Pratt and Whitney Wasp engine, weighed about 4000 pounds loaded, and had a high speed of about 135 mph. Forty BT-1 trainers were converted earlier from Douglas O-2K observation planes and used 420 horsepower Liberty engines.

105.SEVERSKY BT-8 This 1936 Seversky product was the first monoplane basic trainer for the Army and the first specialized type for that role. Thirty aircraft were ordered in 1935 and delivered the next year. Using the typical Seversky elliptical wing platform, a fixed panted landing gear, and enclosed cockpits for the crew of two, the BT-8 used a 450 horsepower Pratt and Whitney Wasp Junior engine and had a two ton gross weight and a maximum speed of 175 mph. The 30 BT-8 airplanes were not considered as successful as the later North American BT-9 series of basic trainers, and there were no follow-up orders.

106.CURTISS WRIGHT CW-19R Designed and built at the St. Louis Division of Curtiss Wright, the CW-19R was a mid-1930s two place military basic trainer and light combat aircraft for the export market powered by either a 350 or 450 horsepower Wright Whirlwind engine. It had a gross weight of 3500 pounds and a high speed of 185 mph. A few were exported to China and Cuba. A single .30 caliber gun could be mounted in the nose; note also the additional .30 caliber gun package mounted on the fairing for the fixed landing gear. The CW-19R was the predecessor design for the CW-22 export trainer and the SNC-1 Navy trainer of World War II.

107.NORTH AMERICAN BT-9 The BT-9 airplane was the standard Army Air Corps basic trainer of the late 1930s, and established North American Aviation Inc. as a premier builder of training airplanes. The BT-9 was derived from the company NA-16 demonstrator of 1935. A total of 266 of these two place fixed gear trainers were delivered between 1936 and 1938 in four variants. The 40 BT-9A versions had a fixed and a flexible machine gun. All were powered by 400 horsepower Wright Whirlwind engines, weighed about 4500 pounds, and had a maximum speed of 170 mph.

108.VULTEE BT-13A A handsome flight photo of the Vultee Valiant basic trainer, the aircraft used to train U.S. World War II military and naval pilots along with its Navy equivalent, the SNV-1 and -2. The Valiant was ordered in 1940 after 1939 tests of a Vultee Model 54 demonstrator, and a total of 7832 were produced between 1940 and 1942 in three variants using a Pratt and Whitney Wasp Junior engine of 450 horsepower. In addition, in 1941 1693 BT-15 models were manufactured with an alternate Wright Whirlwind engine of the same power. With a gross weight of about 4500 pounds, a wingspan of 42 feet, and high speed of approximately 160 mph, the low wing fixed gear "Vibrator" was well regarded by most U.S. pilot trainees.

109.NORTH AMERICAN BT-14 A dramatic formation flight view of BT-14 aircraft shows the next version of the North American basic trainer after the BT-9 series. The BT-14 was a 1940 improved version of the BT-9 with a metal-covered fuselage, a new tail shape, and a revised wing. This North American trainer lost out in production numbers to the Vultee aircraft; 251 BT-14s were produced for the Army in 1940. They used the same Pratt and Whitney Wasp Junior engine as the BT-13, and had a maximum speed of 180 mph.

CARGO

Army cargo transport aircraft were originally designated under a T for Transport category in the early 1920s, however, only four aircraft came under that heading, two of them being nine place Dutch Fokker F-IV single engine monoplanes of 1922 labeled as T-2s. One of these was a specially modified transcontinental record-breaker that currently resides in the National Air and Space Museum in Washington, D.C.

The C for Cargo designation system was initiated with the Douglas C-1 of 1925, a 120 miles per hour single engine biplane with pilots outside up front and eight to nine passenger places inside the cabin. Twenty-six C-1 aircraft were purchased in 1925-27. In the later 1920s, aside from C-1s, only 20 cargo types were purchased by the Army, most of these being Ford or Fokker trimotors slightly revised for military service. In the early 1930s only one aircraft type received a production order, this when 20 Fokker F-14s were purchased in 1931. The Army was unable to fund development of a special cargo plane, so a great variety of aircraft were procured representing many different civil types, such as one three place Curtiss Robin (XC-10), several seven place Fairchild 71 monoplanes (C-8), a few Consolidated four place Fleetsters (C-11 and C-22), a couple of Lockheed Vegas (C-12 and C-17), along with four American Pilgrim monoplanes (C-24). In 1933-34 fourteen Bellanca Airbus (C-27) airplanes with high monoplane wings supplemented by the typical Bellanca lift struts constituted the Army's heavy lift capabilities.

As the 1930s wore on, twin engined low wing all metal civil Lockheed Model 10 and 12 Electras (about 50) and Douglas DC-2s (about 80) purchased under various C designations were procured, thus bringing a few modern transports into the Army fold. These aircraft delivered supplies and spares, including engines and propellers, from Army depots to using squadrons, transported military personnel, and in some cases even delivered engines from the engine manufacturer to airframe manufacturers of new Army aircraft as GFE (Government Furnished Equipment).

With the advent of World War II orders for new Army transport planes poured in to aircraft manufacturers and a production extravaganza ensued. Through the war years all kinds of transports were needed for personnel and cargo. Pre-eminent among the larger aircraft were three types, the twin-engine Douglas DC-3 offshoot Skytrains and Skytroopers and some drafted from the civil airlines in various designations (C-47 through C-53) with about 9700 aircraft involved, the twin-engine Curtiss C-46 Commando totaling about 3200 airplanes, and the 950 four-engine C-54 Skymasters along with 280 Consolidated C-87 transport variants of the B-24 bomber. These aircraft ranged worldwide in support of global military operations. Medium light twin-engine transports used in large numbers included 1400 Beech military C-45 versions of the company civil Model 18, 425 Lockheed C-56 and C-60 versions of their civil Model 18, and over 3400 Cessna UC-78 military versions of the civil Model T-50. These aircraft were supplemented by literally hundreds (about 2200) of single-engine militarized versions of small four to eight place civilian types, Stinsons, Howards, Wacos, Noorduyns, Beech, Fairchilds, and so on, many drafted into service early in the war. Even a 1928 Hamilton monoplane went into the Army as a UC-89 in 1942! This was the wartime air transport fleet of the USAAF.

With the war over new military transport aircraft became more specialized for military missions. Attention was paid to palletizing loads and special cargo loading and tiedown systems were devised. Air drop systems were employed for efficient deployment of both airborne troops and cargo. Short field performance, rough field accomodation, and quick turn-around characteristics were emphasized in aircraft design. A big test of military airlift capability came with the advent of the Berlin Airlift.

The Fairchild C-82 of 1946 and C-119 of 1949 were cargo aircraft designed to allow unobstructed rear end loading. In the early 1950s military versions of large civil airliners like the Douglas DC-6 (C-118) and Lockheed 1049 Constellation (C-121) were ordered in quantity; some of the C-121s as radar picket aircraft. In addition, hundreds of Boeing C-97 versions of the civil 377 Stratocruiser served as cargo planes or tankers. At the same time the Fairchild C-123 twin-engine assault transport was purchased, and the very large four-engine nose-loading Douglas C-124 (34 tons of cargo or 200 passengers) was used by many branches of the USAF. In the late 1950s a new turboprop transport destined to become familiar the world over (even today) was introduced, the Lockheed C-130 Hercules. Less successful over time was the huge Douglas C-133 turboprop.

Another very successful modern cargo airplane introducing the jet engine to the cargo fleet, primarily as a flight refueler for Strategic Air Command bombers, was the C-135 military version of the civil Boeing 707. Boeing had seen a jet tanker requirement by the USAF when it started its 707 commercial development.

The strategic airlift requirements of the services starting in the late 1960s and continuing to date are being covered by the very large Lockheed C-141 Starlifter and the huge C-5A Galaxy, the latter being designed for straight-through loading and unloading, via both nose and tail, of truly outsize items. The latest USAF cargo transport aircraft are the McDonnell Douglas C-10A militarized turbofan DC-10 employed both for cargo and personnel airlift and as an aerial fuel tanker, and the very new McDonnell Douglas C-17A turbofan transport developed purely as a military plane. These

110. NEG301A FORD C-4 One of the earliest Army cargo transport planes, the single Ford C-4 trimotor of 1929 was the equivalent of the civil Model 5A-T all metal airliner, a larger version of the earlier and smaller C-3 type that was the equivalent of a commercial Ford 4-AT. This C-4 model was powered by three 450 horsepower Pratt and Whitney Wasp engines, weighed 13520 pounds fully loaded, and had a maximum speed of 145 mph. The Army later bought four additional slightly revised C-4 airplanes. The C-4 could carry 17 people.

aircraft provide for quick worldwide deployment of U.S. military forces and equipment and for aerial refueling support of smaller aircraft such as fighters to allow tactical air support just about anywhere needed.

In no area have the changes through the last few decades been any more startling than those of the military cargo transport aircraft.

111.SIKORSKY C-6A The Army did not restrict itself to procurement of landplanes as exemplified by this flight photo of a Sikorsky twelve place amphibian aircraft, 11 of which were purchased in 1930 and 1931. These five ton sesquiplanes were just military versions of the civil Sikorsky S-38A. The Army did not have the money to develop new transport designs themselves, and so bought civil aircraft off the shelf, a practice carried out for many years. These airplanes were used for coastal patrol, a mission assigned to the Army. Two Pratt and Whitney Wasp engines of 450 horsepower each powered the C-6A to a maximum speed of 120 mph.

112.Fairchild C-8A The seven place C-8A was a military version of the Fairchild Model 71 civil light transport with the photo showing a twin float seaplane version. Fairchild also made cameras and had considerable interest in aerial photography, as did the Army. For a time C-8 airplanes were designated as F-1 photographic (F for photo) airplanes. A total of 15 aircraft of this type were purchased in 1929-1931. A conventional high wing cabin monoplane typical of the period, the C-8A landplane was a 5500 pound gross weight aircraft powered by a 450 horsepower Pratt and Whitney Wasp providing a high speed of about 140 mph.

112-A.FORD C-9 Another military version of the civil Ford 4-AT trimotor, actually a 14 place C-3A, all seven of which were modified in 1929 by changing to an uprated version of the Wright Whirlwind engine, thereby giving 300 horsepower instead of 235 horsepower each. The Ford weighed five tons, spanned 74 feet, and could speed along at a maximum of 130 mph. The corrugated metal skin of the tri-motor shows up in the photo; the airplane was affectionately called the Tin Goose.

113.FOKKER C-15A The Army in 1931 also bought 20 military versions of the Fokker F-14 six to seven place light airliner, distinguished by a high wooden parasol wing of 59 foot span strut-braced off the fuselage. The aircraft shown, the C-15A, was a C-14 converted in 1932 with a new Wright R-1820 Cyclone engine of 575 horsepower. This airplane was set up as a flying ambulance capable of taking four litters in its fuselage, which was painted white. The red cross is shown on the fuselage side in the photo.

114.NORTHROP Y1C-19 Three military versions of the civil Northrop Alpha low wing monoplane were purchased by the Army in 1931 and 1932, another case of the service buying off the shelf. The pilot sat outside in an open cockpit and four passengers sat inside the fuselage behind the single engine. The all-metal aircraft had a cantilever low wing and fixed landing gear. A 450 horsepower Pratt and Whitney Wasp powered the 4500 pound airplane to a 158 mph high speed. The Alpha was a breakthrough in metal aircraft design at the time.

115.LOCKHEED Y1C-23 A single example of a civil Lockheed 8-D Altair light transport was obtained by the Army in 1932. A very modern design for its day, this two place Altair had a metal fuselage and enclosed cockpits along with a wooden low wing of 43 foot span into which the main landing gear retracted. Based in Washington, D.C., the Lockheed served as a fast VIP transport with a high speed over 200 mph. The ubiquitous 450 horsepower Pratt and Whitney Wasp radial engine was up front. Another Altair, the Y1C-25, was purchased the next year; it was similar to the Y1C-23 except for a fuselage of standard wood construction.

116.AMERICAN Y1C-24 Another Army purchase in 1932 during "The Great Depression" was the civil American Pilgrim 100-B, a large (57 foot span) ten place high wing monoplane of then-conventional design and construction; wing and horizontal tail were strut-braced, and the landing gear was fixed. The pilot sat high and forward atop the wood fuselage. Four Pilgrims were bought. A single nine cylinder Wright Cyclone radial engine of 575 horsepower provided the 7000 pound aircraft a maximum speed of about 135 mph. The Pilgrims proved to be reliable light utility transports.

117.BELLANCA Y1C-27 Yet another example of the Army purchasing a civil aircraft as a cargo transport was the single engine Bellanca P-300 Airbus of 1933-34, 14 of which went into service as Y1C-27s and C-27As, all of which were later updated with more weight and more powerful engines as C-27Bs and C-27Cs. A typically unique Bellanca high wing 65 foot span monoplane, with lift struts connecting wing, fixed landing gear, and fuselage, the 15-place Airbus was the heavy (11000 pound gross weight) transport of the day. The new Cyclone engine was rated at 750 horsepower; high speed was 143 mph.

118.CURTISS YC-30 During the depression, the St.Louis Division of the Curtiss Wright Corporation took on the task of developing a large low cost commercial airliner; the result was the twin-engined T-32 Condor biplane whose only bow to advanced design was a retractable landing gear. Along with major civil airlines and the U.S. Navy, the Army purchased two Condors as YC-30s in 1934 with places for 19 occupants. One Condor was modified for presidential use, but was never employed in that capacity. Wingspan was 82 feet, and two Wright Cyclone engines each of 650 horsepower could power the 18500 pound transport to a high speed of about 160 mph.

119.LOCKHEED XC-35 Ordered in 1936 and delivered the next year, the single XC-35 was a Lockheed 10 civil transport modified for Army research into aircraft structures and systems suitable for high altitude flight. Changes included a new fuselage of circular cross-section adaptable to pressurization, a cabin pressurization system, and 550 horsepower Pratt and Whitney Wasp engines equipped with General Electric turbosuperchargers. The five place aircraft had a 55 foot wingspan, a gross weight of 10500 pounds, and a maximum speed at altitude of 240 mph. The XC-35 paved the way for future high altitude Army aircraft designs.

120.LOCKHEED Y1C-36 The photo shows one of three military equivalents of the Lockheed 10A Electra commercial transport bought by the Army for personnel transport in 1937. A ten place twin engine aircraft, the Electra was a five ton machine powered by Pratt and Whitney Wasp Junior engines of 450 horsepower, each providing a top speed of just over 200 mph. With a World War on for the U.S. in 1942, small batches of civil Electras were impressed into Army service. These included a mixture of Model 10As, 10Bs, and 10Es totaling 26 airplanes, with various Pratt and Whitney Wasp, Wasp Junior, and Wright Whirlwind powerplants.

121.DOUGLAS C-38 Representative of several Army variations of the Douglas DC-2 civil airliner with Wright Cyclone engines, this C-38 of 1937 was a C-33 with a DC-3 tail, sometimes called a DC-2 and 1/2. The Army was most interested in the revolutionary new Douglas airliner and purchased one in 1936 as an XC-32, and 24 more were drafted for war. Twenty similar Douglas C-33 and C-34 16 place aircraft were purchased in 1936-37, and 35 C-39s came into the inventory in 1939. These Douglas aircraft spanned 85 feet, weighed 18500 pounds, had a maximum speed of about 210 mph, and cruised at about 155 mph.

122. BEECH UC-43 The company photo shows yet another civil aircraft design first obtained by the Army in the initial year of World War II and followed by many more in 1942. Three civil Beech D17S Staggerwing biplanes were purchased as YC-43s, followed in 1942 by a production order for 207 more UC-43 Traveler five place light utility transports. A bevy of various civil Staggerwings were also impressed into military service that year (totaling 118 aircraft) and designated from UC-43A through UC-43K. The UC-43 had a gross weight of 4800 pounds, spanned 32 feet, and used a Pratt and Whitney Wasp Junior engine of 450 horsepower, giving a top speed of 195 mph. These Beech aircraft were easily distinguished by the reverse stagger of the biplane wings.

123.BEECH C-45 The C-45, later labeled the Expeditor, was an Army version of the Beech 18 twin-engined civil light transport. The aircraft in the photo has pre-World War II markings, one of 11 six place transports ordered and delivered by Beech in 1940. They were later redesignated as UC-45s, the U denoting utility missions. Powered by Pratt and Whitney Wasp Juniors each of 450 horsepower, the C-45 could fly to a maximum speed of 225 mph.

124.BEECH C-45G One of the final models (rejuvenated postwar) of the Beech Expeditor is shown in a manufacturer's flight photo. A total of 1390 Beech Expeditors were used during wartime by the Army, and were labeled from C-45A to UC-45F with minor differences. Most all, 1137 aircraft, were the seven place C-45F models. The G and H versions were post-war re-manufactured airplanes and were used for years afterwards. Engines and performance were similar to the early versions.

125.CURTISS C-46A The prototype of the Army C-46 Commando was intended as a CW-20 36 place pressurized civil transport and used two 1700 horsepower Wright R-2600 engines, but the onset of World War II incited Army interest in the large cargo carrying potential of the aircraft. In 1941 a 25 plane order was placed for a revised C-46 version using two 2000 horsepower Pratt and Whitney R-2800 Double Wasps. The C-46A Commando cargo transport was then selected for mass production with first deliveries coming in 1942. Three Curtiss factories cooperated in producing a total of 1483 C-46As; two aircraft to be built by Higgins in New Orleans were canceled. The C-46 type was most famous for carrying supplies over the "Hump" from India to China during World War II.

126.CURTISS C-46A The first C-46A-50 Commando is shown in this beautiful company flight photo. It was also the first aircraft without camouflage paint, thereby saving considerable weight. The C-46A weighed 49600 pounds normally loaded, or ten tons more than a C-47. With a span of 108 feet it was the largest twin-engined transport in U.S. service. The C-46 was a standard USAF military aircraft for many years after the war and served in Korea. Many were also sold off as surplus and later used as commercial cargo carriers.

127.CURTISS C-46D The other high production Commando, shown at low altitude as if ready to drop paratroopers, was the C-46D version shown in this March 1945 Curtiss photo. A total of 1410 D models were put out by Curtiss in 1943-44 with a revised nose and a double door arrangement. The 26 ton C-46D had a high speed of 240 mph and cruised at about 190 mph. At war's end many C-46 orders were canceled. In addition to the A and D version Commandos, Curtiss produced 234 F variants quite similar to the C-46D.

128.DOUGLAS TC-47B The 24 place C-47 Skytrain was a military version of the commercial Douglas DC-3, often called the "Gooney Bird", and considered one of the most useful weapons of World War II. For Army use the fuselage was strengthened, including the floor, and door size was increased, all to carry military cargo. The two engines were 1200 horsepower Pratt and Whitney Twin Wasps providing a high speed of about 220 mph. The Skytrain was put into production starting in 1941, and 953 C-47s, 4931 C-47As with minor changes, and 3241 C-47Bs having increased engine supercharging with two speed impellers (including 133 TC-47B trainer versions like the aircraft shown in the photo) were manufactured into 1944. The C-47 was a 15 ton gross weight airplane with a 95 foot wingspan.

129.DOUGLAS XC-47C The C-47 was even tested in 1943 using twin Edo floats with wheels incorporated, making it an amphibian with some fuel carried in the floats. The single XC-47C is shown in flight. Floats added over two tons to aircraft weight and reduced high speed by about 30 mph. No real operational use was found for the conversion and development was discontinued.

130.DOUGLAS C-47D Many C-47 aircraft stayed in service long after World War II. The C-47D shown was one of many C-47B aircraft revised to eliminate the extra engine supercharging originally intended to allow flying over the "Hump" in the CBI Theatre during the war. Many other DC-3 types were acquired by the Army , including airliners drafted for service. These were designated variously, such as C-48, C-49, C-50, C-51, C-52, etc. The C-53 variant was called a Skytrooper and, as the name implies, was used mainly as a troop carrier. A total of 377 C-53s were produced.

131.DOUGLAS C-54B The C-54 Skymaster started out in 1941 as a planned DC-4A four engine airliner, but with America's entry into the war late that year the Army took over initial Douglas production of 24 aircraft as C-54s delivered in 1942. In the next year 427 C-54A and C-54B versions like the one shown were produced in two plants. The C-54B was a 50 seat unpressurized military transport with a crew of six and a gross weight of 73000 pounds powered by four Pratt and Whitney R-2000 (bored out Twin Wasp) engines each of 1350 horsepower providing a maximum speed of about 270 mph. It could also be used as a cargo carrier. C-54s were employed in the Korean War and also in the Berlin Airlift.

132.DOUGLAS C-54D The C-54D version of the Skymaster was similar to the B model except for use of a new version of the R-2000 engine. A total of 350 D models were produced at the Douglas Chicago plant in 1943 and 1944. The C-54 was not pressurized, and thus cruising with passengers aboard was normally limited to the lower altitudes. Two more C-54 variants were produced; 75 of a C-54E long range version in 1944, and 76 C-54G models in 1945. The C-54s gave excellent service in World War II and afterwards during the early stages of the Cold War. Some aircraft served through the 1950s. Many others were sold and converted to civil transports.

133.NOORDUYN C-64A The Noorduyn C-64 Norseman was a rugged Canadian bush airplane designed for civil light transport use and purchased during World War II in 1942 and 1943 for military utility duties. The aircraft in the photo is an equivalent Royal Canadian Air Force Norseman VI. After seven YC-64 service test airplanes were procured by the U.S. Army in 1942 their success allowed purchase of 746 C-64A ten place models (eight passengers) the next year. Six C-64B versions with twin floats were also procured. The 7400 pound landplane C-64A used a 600 horsepower Pratt and Whitney Wasp engine and had a high speed of 160 mph.

134.LOCKHEED C-69 The C-69 Army designation covered a total of 20 new pressurized Lockheed Constellation civil four engine low wing monoplane transports after they had been inducted into World War II service. Distinguished by a long airfoil-contoured fuselage and a triple tail, the 69 place 41 ton gross weight C-69 entered Army service in 1945, but the aircraft were returned to the airlines after the war. Four 2200 horsepower Wright 18 cylinder R-3350 radial engines powered the C-69 to a top speed of about 330 mph at altitude. One of the aircraft was a C-69C delivered in 1946.

135.DOUGLAS C-74 The Douglas Globemaster heavy transport was designed during World War II to carry 129 passengers and crew or large cargo items over very long ranges. The first of 14 C-74 transports was delivered right after the war, and the rest came into service in the next two years. With a gross weight of 125000 to 145000 pounds, the C-74 had a wingspan of 173 feet and could carry up to 28 tons of cargo over the shorter ranges. Power was supplied by four Pratt and Whitney R-4360 28 cylinder Wasp Major engines driving four bladed propellers, and a high speed well over 300 mph was obtained. The C-74 participated in the Berlin Airlift.

136.CESSNA UC-78 The UC-78 Bobcat was, like the AT-8 and AT-17, a military version of the commercial Cessna T-50 light transport of 1939, and the Crane shown in the photo was a Canadian trainer equivalent. The U.S. Army accepted the quite astonishing total number of 3428 UC-78, -78A, -78B, and -78C Bobcats in 1942 and 1943. Seventeen of these were impressed civil aircraft labeled UC-78A. The 1806 UC-78B aircraft were painted all silver and had wooden fixed pitch propellers. The Bobcats were four place 5700 pound airplanes using two Jacobs radial engines of 225 horsepower each. High speed was 175 mph.

137.FAIRCHILD C-82A The C-82A Packet was a unique twin boom and tail cargo aircraft designed to allow movement of large military items through clam shell doors at the rear end of the fuselage. After a first flight in September 1944, delivery of a prototype in 1945, and delivery of 100 production aircraft just after World War II ended, a further 1945 order was received for 100 more. A total of 204 Packets were produced before cancellation of contracts. The 21 ton gross weight C-82 used two Pratt and Whitney Double Wasp engines of 2100 horsepower. High speed was about 225 mph.

138.CONVAIR C-87 A transport version of the B-24 Liberator bomber, the C-87 of 1942 was a modified B-24D with provision for a crew of five and 20 passengers. No armament was carried, there was no cabin pressurization, and windows were provided in the fuselage. A total of 282 Liberator transports were delivered, mostly in 1942; six of these were C-87A models with ten sleeper berths. The long range of the Liberator made it an important VIP transport in World War II. Gross weight was 28 tons; cruising and high speeds were about 220 and 305 mph respectively. As with the B-24, power was supplied by four turbosupercharged Pratt and Whitney Twin Wasp engines of 1200 horsepower each.

139.BOEING XC-97 The XC-97 Stratocruiser was a transport version of the B-29 bomber, the wings, powerplants, and tail being the same and the fuselage all new in a pressurized "double bubble" cross-section with rear ramp. It was capable of carrying 134 passengers and crew or cargo and crew. The photo shows one of three prototypes in October of 1944 and emphasizes the high aspect ratio of the wing. Maximum speed with the Wright R-3350 engines was 332 mph at altitude. Six YC-97 versions were also built with the Wright engines before a change to Pratt and Whitney Wasp Major powerplants was made.

140.BOEING KC-97G The final production model of the C-97 series , with a total of 888 Stratofreighters having been built. Most all C-97s were powered by four Pratt and Whitney R-4360 Wasp Majors each of 3500 horsepower and had B-50 bomber tail surfaces. Most all (572) aircraft were KC-97Gs, and were employed as tankers for B-50 bombers using the Boeing Flying Boom refueling system. The last KC-97G was completed in July of 1956, production having taken place over a period of 12 years. Cruising and high speeds were 300 and 375 mph respectively. The KC-97G airplanes were superceded by the KC-135 jet tankers.

141.CONVAIR XC-99 Though only a single aircraft was procured by the USAF, the XC-99 was notable for its size as a transport version of the B-36 intercontinental bomber. The major change from the B-36 was a huge new double deck fuselage to carry 375 to 400 troops or 300 stretchers, or up to 50 tons of cargo. First flight was in late 1947, and the single aircraft was used for ten years on special missions. Wingspan was 210 feet, and gross weights, normal and overload, were 265000 and 320000 pounds respectively. Six Pratt and Whitney R-4360 Wasp Major engines of 3500 horsepower each in a pusher propeller installation (like the B-36) powered the XC-99 to a top speed of 300 mph. Auxiliary jets were never added to the XC-99 like they were to the B-36; there was no need for a dash speed.

142.FAIRCHILD C-119F The Flying Boxcar in its final form; the C-119F and C-119G models had several changes, including an increase in design gross weight to 85000 pounds and boom vertical fins, but a major change was reversion to Wright R-3350 engines rated at 3500 horsepower each. The 69 place Boxcar had a wingspan of 109 feet and a high speed of about 260 mph. About 260 C-119Fs and 480 C-119Gs were produced, the F with Hamilton and the G with Aeroproducts propellers. C-119s served in Korea, were provided to some foreign air forces, and some ended up in Air Force Reserve and Air National Guard squadrons.

143.LOCKHEED C-121C The military Lockheed C-121C Constellation was a USAF version of the commercial L-1049F transport and had an R7V-1 naval counterpart. The C-121C was ordered in 1951 and came into service in 1955. It had a 135000 pound gross weight and was powered by four Wright R-3350 engines each of 3500 horsepower. Maximum speed was about 360 mph. There were many C-121 versions in the 90 total airplanes built. Some were VIP transports; others were reconnaissance and airborne early warning models with radomes such as the RC-121C and the EC-121C aircraft, the latter equivalent to the naval WV-2. Over 40 EC-121D aircraft came out in 1955 using wingtip fuel tanks, a 70 ton gross weight, and Wright Turbo Compound engines; there the Navy equivalent was the WV-3.

144.FAIRCHILD C-123K The C-123 Provider twin engine cargo and assault transport aircraft with a rear ramp was originally a design of Chase aircraft and later Kaiser-Fraser Corporation, but revisions and production were taken over in 1953 by Fairchild, which built 300 C-123Bs with most deliveries made in 1955. A unique feature was jettisonable internal nacelle fuel tanks, and many Bs were fitted with wing drop tanks. Powered by two Pratt and Whitney Double Wasp engines each of 2500 horsepower, the C-123B had a top speed of 240 mph and a maximum takeoff weight of 57800 pounds. The C-123J and K versions (the latter shown) were equipped with two Fairchild J-44 auxiliary jet engines each of 1000 pounds thrust. Maximum gross weight was 30 tons. The C-123s were used in Vietnam, being noted chiefly for employment on defoliation missions.

145.FAIRCHILD C-123K Flight photo of a C-123K shows the aircraft with the underslung wing jet pods but without wing drop tanks. The jet engines were used primarily for takeoff under full load conditions. First flight of a C-123 was in December of 1952 and the first Fairchild model acceptance was in July of 1954. An updated C-123J or K carried a crew of three and 58 troops, or it could carry 46 litters and ten attendants for medical evacuation. At a 52250 pound takeoff weight it could carry 5000 pounds of cargo for a combat radius of 500 nautical miles. C-123s were rugged, long-lived aircraft.

146.DOUGLAS C-124C The C-124 heavy cargo transport was labeled the Globemaster II since it was a C-74 Globemaster with a new fuselage of greater capacity and with more loading facilities. Wings, tail, and powerplant were like the C-74. The main cabin had a 10000 cubic foot capacity, and could house heavy equipment or 127 level litters. There was a crew of five with a relief crew of three for long flights. First flight of the initial C-124A version, of which 204 were built, was in April 1950. Production of the A was completed in January 1953. The C-124C first flew in February 1953, and production of 243 of this model ended in April of 1955.

147.DOUGLAS C-124C The C-124C Globemaster II was like the C-124A except for a new version of the engine and new propellers. The photo shows the nose loading feature of the C-124 with ramps and clam shell doors. There was also an electrically operated elevator in the floor just aft of the wing. The C-124C was powered by four Pratt and Whitney R-4360 Wasp Major engines each of 3800 horsepower. Wingspan was 174 feet and design gross weight 185000 pounds, with an overload capability to 216400 pounds. Maximum speed was 270 mph with cruise speed about 230 mph. The C-124 served the USAF into the 1960s.

148.LOCKHEED C-130A Probably no military transport in the world is as well known as the Lockheed C-130 Hercules four engine turboprop transport. The photo shows an updated C-130A of an Air Force Reserve squadron. Hundreds of C-130 aircraft in many versions have been employed in a wide array of transport missions, and they have served with the air forces of many nations. The original C-130A was accepted in September 1955 and went into service in December 1956. A rear ramp loading transport with compartment pressurization, the C-130A had a crew of four and a capacity for a maximum of 92 troops. The photo shows underwing fuel tanks retrofitted to early models after being employed first on the C-130E models.

149.LOCKHEED C-130A This striking head-on view of a C-130 shows the cleanliness of the basic design and the relatively slim frontal area of the four Allison T-56 3750 horsepower (plus 740 pounds of jet thrust) turboprop engines. The external fuel tanks are outboard of the engines; internal fuel is in the wings. With a wingspan of 132.6 feet and a maximum gross weight of 135000 pounds (C-130B) or 155000 pounds (C-130E), the C-130B had a high speed of about 380 mph. It could carry a 25000 pound design payload for a 2800 nautical mile range. Maximum overload payload was 50000 pounds carried for a 750 nautical mile range.

150.LOCKHEED AC-130 The C-130 Hercules was converted to many configurations for special missions. The photo shows some of the armament aboard a "Puff, the Magic Dragon" attack version. The near weapons are high rate of fire multi-barrel machine guns. Other guns used were 20 and 40mm, and sometimes even a 75mm cannon installation. The aircraft flew in a circle over and around a ground target and literally poured out gunfire from the inboard side of the aircraft.

151.CONVAIR C-131A The C-131A Samaritan shown in the photo was one of 26 USAF medical evacuation transport versions of the civilian Convair 240 postwar airliner and was quite similar to the T-29A trainer. Ordered in 1952 with deliveries completed in 1954, the C-131A could transport 27 litter patients. A trim 43000 pound gross weight twin-engined piston-powered transport with Pratt and Whitney Double Wasp engines each of 2500 horsepower, the C-131A had a maximum speed of 275 mph and cruised at about 250 mph. Later C-131 models included B, C, D, and E variants.

153.DOUGLAS C-133A The C-133 Cargomaster was designed as a heavy lift military transport right from the start in 1952 as a turboprop aircraft to achieve longer ranges than jet engine powered aircraft of the day could deliver. A high wing design with a pressurized cabin and rear ramp loading, this very large (180 foot span) transport was powered by four Pratt and Whitney T34 turboprop engines, each delivering 6500 horsepower plus 1250 pounds of jet thrust and driving three-blade 18 foot diameter Curtiss Turboelectric propellers. A total of 34 airplanes were delivered starting in August 1957. Normal maximum takeoff gross weight was 275000 pounds. Among other large cargo items the C-133 could transport USAF ballistic missiles such as Atlas, Thor, and Titan.

152.DOUGLAS C-133A Flight photo of a C-133A Cargomaster shows the long fuselage providing carrying capacity for many outsize objects. The length of the main cabin was over 97 feet, and the cargo bay cross-section was 12 feet by 12 feet. Normal crew was four—pilot, copilot, systems engineer, and navigator. In addition to the C-133As, 15 C-133B aircraft were produced starting in late 1959; these versions had engines uprated to 6950 shaft horsepower plus 1375 pounds of jet thrust each and allowable normal maximum gross weight went to 286000 pounds with overload takeoff going to 300000 pounds. At the higher weight a 60000 pound payload could be carried for a range of 3500 nautical miles.

154.12BOEING KC-135A The KC-135 Stratotanker military jet transport was designed for aerial refueling the USAF jet bombers. The earlier KC-97 tankers did not have the refueling speed to keep up. Boeing had a USAF tanker contract in mind when it developed the civil 707 transport, and the KC-135A with Pratt and Whitney J57 jets was similar in design and size to the 707 prototype with the Boeing Flying Boom refueling system added. The KC-135A started operations in 1957. More than 600 tanker aircraft were delivered through 1961. The C-135 was made in several versions, including C-135A and B transports, the latter using four Pratt and Whitney TF-33 turbofan engines each of 17000 pounds of thrust. Some RC-135s were converted for reconnaissance. The KC-135A had a high speed of 585 mph at 30000 feet and cruised at about 520 mph.

155.LOCKHEED VC-140B The C-140 Jetstar was a military version of the Lockheed Model 1329 civil corporate and private jet aircraft. Five C-140A models were delivered to the USAF in 1962 and were used for inspection of world-wide navigation aids. The C-140B, eleven of which were delivered in 1961, was a special mission transport version with a crew of three and either eight or 13 passengers. The C-140 was powered with four Pratt and Whitney JT12 jet engines each developing 3300 pounds of thrust and equipped with thrust reversers. The pressurized aircraft had a maximum takeoff weight of 21 tons, a wingspan of 54 feet five inches, and a maximum speed of 570 mph at 23000 feet.

156.LOCKHEED C-141A The C-141A Starlifter is a very large (160 foot wingspan) four engine long range strategic transport, 284 of which had been ordered by the end of 1966. First flight was in December of 1963 and first delivery in late 1964. These aircraft are still in very active use by the USAAF Military Airlift Command. Powered by four Pratt and Whitney TF33 turbofan engines each of 21000 pounds thrust, giving a maximum speed of 570 mph at 25000 feet and maximum range cruise speed of 506 mph, the C-141A has a range with 34 ton payload of 4200 miles and a range with a 16 ton load of 6160 miles.

157.LOCKHEED C-141A The flight photo of the Starlifter shows the high swept wing with underslung turbofan powerplants and the tee horizontal tail. The aircraft is loaded via a rear ramp and clamshell doors. The pressurized cabin is 70 feet long, over 10 feet high, and nine feet wide, and can accommodate a Minuteman ICBM. Maximum takeoff weight with the Minuteman assembly on board is 150 tons. As a personnel transport, the C-141A has a flight crew of four and can carry up to 154 troops.

158.LOCKHEED C-5A A huge aircraft by any standards, the C-5A Galaxy logistics transport is used by the USAF Military Airlift Command primarily to transport outsize cargo payloads at ranges up to 8000 miles. The C-5A aircraft is the result of a 1964 airframe design competition won by Lockheed-Georgia (with Boeing and Douglas as losers) and an engine competition won by General Electric with their GE1/6 high bypass ratio turbofan. Of a configuration generally similar to the earlier C-141, but much larger, the C-5A has a wingspan of 223 feet and a maximum takeoff gross weight of 764500 pounds at a 2.25 g load factor. Maximum speed is 570 mph and average cruise speed 507 mph. A straight through loading and unloading arrangement was designed in with both nose and tail access provided.

159.BEECH VC-6A The VC-6 was a military version of the Beech civil King Air 90 used for light personnel transport and special mission duties in the middle 1960s. It was a twin turboprop six to 10 seat pressurized aircraft powered by Pratt and Whitney PT-6A engines of 550 equivalent shaft horsepower, each with full-feathering Hartzell propellers. Wingspan was 50 feet, maximum takeoff weight 9650 pounds, and maximum speed 270 mph at 21000 feet.

160.DEHAVILLAND C-7A The C-7A light twin engine military assault transport was built in Canada as the DHC-4A and supplied to several countries, including the U.S., as the Caribou. The prototype first flew in July 1958 and five aircraft were provided to the U.S. Army for evaluation as YAC-1 transports. Later, 159 planes were ordered by the Army as the CV-2 (was AC-1). In early 1967 the remaining 134 transports were transferred to the USAF and redesignated C-7A. The aircraft was an unpressurized rear ramp loading high wing transport of 14 tons gross weight powered by Pratt and Whitney R-2000 piston engines each of 1450 horsepower, thus providing a high speed of 216 mph and a cruise speed of 154 to 182 mph. The aircraft could carry 32 troops.

161.DEHAVILLAND YAC-1 A manufacturer's flight photo dated August 1959 of one of the first YAC-1 Caribou transports ready for evaluation by the U.S. Army. The sharp upswing of the fuselage made provision for a simple rear ramp design. The Caribou was the biggest aircraft in the Army inventory at the time. Testing resulted in a large production order. A typical cargo load could be two fully loaded jeeps. Maximum payload was 8700 pounds and the range carrying this load was 242 miles.

162.DEHAVILLAND C-8A A follow-on design to the C-7A Caribou, the C-8A Buffalo won a U.S. Army design competition for a STOL tactical transport in 1962. The new high wing aircraft had a larger fuselage still using rear ramp loading and adding turboprop engines. The Buffalo first flew in April 1964 as an Army YAC-2. In January 1967 the USAF took over all Army transport aircraft and the Buffalo became the C-8A. The airplane has a wingspan of 96 feet, a maximum gross weight of 49200 pounds, and is powered by two General Electric T64 turboprop engines of 2850 equivalent shaft horsepower. Accomodation is made for a crew of three and 41 troops, 35 parachute troops, 24 stretchers and six attendants, or cargo.

163.MCDONNELL DOUGLAS C-9A The C-9A Nightingale is the USAF version of the commercial DC-9 Series 30 jet transport arranged as a medical evacuation aircraft. The first C-9A appeared in June of 1968, and 21 aircraft had been delivered by early 1973. The Nightingale can transport 30 to 40 litter patients, along with five nurses and technicians. Three hydraulically-actuated entrances facilitate litter conveyance. The aircraft is powered by two Pratt and Whitney JT8D-9 turbofan engines, each of 14500 pounds thrust. Wingspan is 93 feet five inches, maximum takeoff weight 49 tons, and maximum cruise speed 565 mph.

164.MCDONNELL DOUGLAS C-10A This USAAF flight refueling and cargo aircraft is a military derivative of the DC-10 civil transport first flown in July, 1980. A crew of three are normally on the flight deck with various arrangements possible in the large cabin. The Boeing flying boom aerial refueling system is installed. The aircraft is powered by three General Electric CF6 turbofan engines and is 182 feet long. Design fuel capacity is 356000 pounds, maximum cargo payloads of up to 164000 pounds can be carried, and design maximum takeoff weight is 590000 pounds. With the maximum cargo load the range is 4370 miles, and maximum ferry range without refueling is 11500 miles.

165.BEECH C-12A The C-12A is a military version of the civil Beech Super King Air 200 light twin turboprop transport with places for 15 people, including the pilot. A sleek, pressurized aircraft with a tee tail powered by two 850 horsepower Pratt and Whitney PT-6 turboprop engines, the 12600 pound gross weight light transport has a high speed capability of 333 mph at 15000 feet. The aircraft first flew as a civil transport in October of 1972. It has basically the same fuselage as the King Air 100, but includes an increased wingspan of 54 feet six inches, a new tail, more powerful engines, and more fuel, along with increased weight.

166.BEECH C-12A The C-12A was first delivered to the military in 1974-75 when the U.S. Army purchased 20 aircraft and the USAF bought 14. In 1976 and 1977 the Army added 40 additional C-12 aircraft and the USAF added 16. In addition, the Navy ordered 22 C-12s. Some of the Army aircraft have been modified as RU-21Js with special mission avionics under a Cefly Lancer program. By early 1978 a total of 113 C-12As were being operated by the Army and the Air Force.

167.MCDONNELL DOUGLAS C-17A The C-17A Globemaster III is a very large strategic transport aircraft designed specifically for the role. Contracted in August 1981 to increase the nation's strategic airlift capability, the C-17A program was in serious trouble in 1993, but President Clinton has decided to go ahead with the aircraft, and 24 had been produced by early 1996. In June 1996 the USAF signed contracts for $16.2 billion with McDonnell Douglas to buy 80 more C-17As to be built over seven years, which will eventually bring the fleet to 120 airplanes. First flight took place in May of 1992. Powered by four Pratt and Whitney F117 turbofans each of 40700 pounds thrust, the aircraft can carry many outsized items, including a main battle tank. Wingspan is 165 feet, and the cargo compartment is 88 feet long by 18 feet wide by 12 feet four inches high. Maximum takeoff gross weight is 580000 pounds, and a normal cruise speed would be Mach .77 at 28000 feet. Range with 150000 pounds of payload is 3110 miles. In June 1996 the USAF signed contracts for $16.2 billion with McDonell Douglas to buy 80 more C-17As to be built over seven years. This will eventually bring the C-17A fleet to 120 airplanes.

ELECTRONIC WARFARE

In recent years electronic warfare aircraft have assumed a role of almost over-riding importance in modern war. They carry electronics/avionics systems to sense enemy force dispositions at long range and provide quick real-time information to battle commanders so friendly forces can be properly concentrated and dispatched to defeat the enemy. Information sensed through optics, infra-red, and radar means can be quickly processed by computer and presented in optimum form, enabling realistic tactical decisions to be made by force commanders. Electronic warfare aircraft can be used for surveillance or command and control in or near a combat area.

The initial aircraft designated in the specific E- series were U.S. Navy types, first the piston-engined Grumman E-1 carrier-based airborne early warning airplane, and later the E-2 twin turboprop airplane with first deliveries in 1964 to replace the E-1.

Before use of the primary E- designator USAF airplanes employed for electronic warfare were also specially-converted versions of large military cargo aircraft, these having the space for installation of many specialized electronics operator crew stations and large antennas and radomes. Notable examples were the Lockheed EC-121 and RC-121 piston-engined Constellations, Lockheed EC-130 turboprop Hercules, and the RC-135 turbofan version of the Boeing 707. The first RC-135s were ten aircraft converted in 1965 from KC-135 tankers to RC-135B reconnaissance aircraft. From then on well into the 1970s many other C-135 aircraft were converted for a reconnaissance mission with various combinations of electronics systems.

In 1977-78 the new Boeing E-3 Sentry, an AWACS version of the 707, was introduced into service with its large external radome above the aft fuselage providing surveillance and control of airspace for well over 200 miles in every direction. This capability was effectively used in the Gulf War of 1991.

The next electronic warfare aircraft type in the series was the Boeing E-4A Advanced Airborne Command Post airplane built to serve as a platform for a battle staff monitoring war situations and effecting strategic military command and control from the air in event of enemy destruction of similar ground based facilities. This platform is the E-4 modification of the civil Boeing 747 and is a replacement for the EC-135.

An aircraft somewhat similar to the E-3A Sentry is the E-8A Joint Surveillance Tactical Attack Radar System (J-STARS) specifically designed to monitor the movement of ground force traffic using a down-looking radar. Two prototype E-8 aircraft were tested and then employed in the Gulf War in 1991 with considerable success, providing target information to friendly force artillery.

168.M-5 BOEING E-3A The E-3A Sentry Airborne Warning and Control System (AWACS) aircraft, delivery of which started to the USAF in March 1977, are special military versions of the Boeing 707-320 civil transport equipped with an extensive array of electronic equipment for airborne early warning and a command center for identification, surveillance, and tracking of enemy air forces as well as friendly forces. The most notable externally mounted equipment is the fuselage strut-supported Westinghouse downward-looking surveillance radar within a 30 foot diameter rotating circular radome. The E-3A is powered by four Pratt and Whitney TF33 turbofan engines each rated at 21000 pounds of thrust. Maximum takeoff gross weight is 325000 pounds. Twenty-four E-3As were built. Improved E-3B and E-3C versions have updated electronics. Additional E-3As have gone to NATO, England, France, and Saudi Arabia.

168-A.Boeing E-4A The Advanced Airborne Command Post airplane is a military version of the civil Boeing 747B transport employed as a command center to direct military operations from the air in event of a devastating attack upon the US. Command and control of a flexible American response would be exercised by a battle staff aboard the E-4A, several of which were ordered from Boeing. The first two aircraft were ordered in February 1973 and delivered late in that year. The E-4A replaced the earlier EC-135 Airborne Command Post airplanes. Characteristics and performance of this aircraft are similar to those of the civil 747.

PHOTO

Photographic reconnaissance could be accomplished by Army observation planes to a certain extent, as long as hand-held cameras could be used by the rear seat man, however, in 1930 the Army started a specialized F- series of designations for aircraft with the single mission of aerial photography.

Perhaps appropriately, since the Fairchild company was a major developer of aerial camera equipment, the Fairchild commercial F-71 high wing cabin monoplane, the three place C-8 in the Army, was designated as the F-1 photo plane and 14 aircraft were procured in 1930-31. It took nine more years for the Army to order another plane with the primary mission of photography, this the Beech F-2 version of the civil B18S light two-engine transport. In 1940 14 F-2s were bought, and later 55 C-45s were converted into F-2As and F-2Bs. The F-2 had two large vertical mapping cameras installed in the belly of the cabin, and a special hatch in the cabin door could accommodate a large camera for high altitude oblique photography with the crew using the oxygen system installed in the F-2. The F-2 did not have the performance required for photo reconnaissance in combat areas so, in World War II, photo versions of combat aircraft were employed for this mission. The F-3 designation was given to a few Douglas Havoc types converted for a camera installation aft in the belly.

When the U.S. entered World War II the need for a fast high altitude photo airplane was recognized, and the Lockheed P-38 fighter was selected for conversion to 119 F-4 and F-5 photo versions; the first F-4 was produced in December, 1941. By April of 1942 F-4s were performing operational missions in the Southwest Pacific. A batch of F-5A versions were started into production as early as January of that year. In England the 8th Air Force started reconnaissance missions over Europe in early 1943 at 25000 feet altitude, but losses became high. F-5s were in almost all theatres of operation. They mapped the French coast prior to the D-day invasion and also mapped a good portion of Italy. As an example of an F-5 camera installation, the F-5B had photo equipment in the nose compartments replacing guns. A six inch K-17 chart camera was in the forward compartment, two of these were in the center compartment, or one 12 inch or 24 inch reconnaissance camera, and the aft compartment held two 24 inch K-17 or one 24 inch K-18 reconnaissance camera. There were F-5 variants of all the P-38 versions; the final F-5G was a modified P-38L.

The other World War II fighter modified for photo reconnaissance was the P-51 Mustang. From 1942 on almost 500 various Mustang versions were converted to that mission as F-6 aircraft with fuselage cameras.

In addition, bombers were converted for long range photo reconnaissance including F-7 variants of the B-24 with as many as 11 cameras, and F-9s as converted B-17F and B-17G camera planes. The F-13 was an armed photo version of the Boeing B-29 with several cameras, and over a hundred of these conversions were used extensively in the Pacific late in the war.

After the war was over, in 1947, in common with other types, the F-13s were redesignated as RB-29s for reconnaissance bomber instead of photo plane, and these were employed during the Korean War.

There were several experimental prototype F- series aircraft, such as the Hughes XF-11 and the Republic F-12 Rainbow. A Northrop F-15 Reporter version of the P-61 fighter saw very limited production, but at the end of World War II the F- series designator standing for photo (or Foto) was obsolete. The reconnaissance modifier R to the basic designation of an aircraft, such as RB-29, allowed a much broader definition of snooping activity, using not only photography, but radar mapping along with infrared or other sensor inputs, including the human eyeball.

169.BEECH F-2 Another military version of the Beech 18 light twin-engined transport, the F-2 was a special Army photographic reconnaissance aircraft modified from a commercial B-18S design. Fourteen three place F-2s were ordered and delivered in 1940 with cameras in the fuselage belly. Powered with two Pratt and Whitney Wasp Junior engines each of 375 horsepower, the F-2 weighed 7725 pounds loaded and had a high speed of 225 mph. During World War II, 13 heavier F-2A aircraft were modified from UC-45s and 42 F-2B versions from C-45Fs, the latter with a trimetragon camera installed.

170.LOCKHEED F-5A An example of a World War II first line fighter aircraft modified for photo reconnaissance, the F-5A Lightning was one of several P-38 versions, this being one of 181 P-38Gs so modified in 1942 with cameras in the nose. Using two turbosupercharged Allison V-1710 engines of 1325 horsepower each, the eight ton gross weight F-5A could achieve a high speed of 390 mph at altitude. Hundreds of unarmed Lockheed F-4 and F-5 photo planes were used very successfully throughout World War II in many theatres of operation, including photo-mapping of most of Italy.

171.BOEING F-13A In 1944 and 1945 some B-29A and B-29B Superfortresses were converted for photo reconnaissance and redesignated F-13A; the photo shows one of these aircraft. Bomb bay fuel tanks were added to provide extra range, and multiple cameras were installed. A crew of nine was normally aboard. After the war in 1948 the few F-13As left were redesignated as RB-29As, the R for reconnaissance. Like the bomber version, the F-13As were powered by four Wright R-3350 engines of 2200 horsepower each, weighed about 141000 pounds loaded, and had a maximum speed of about 360 mph.

172.NORTHROP F-15A A reconnaissance version of the P-61 Army Black Widow night fighter, the F-15A Reporter design was initiated in 1945 with conversion of one XP-61E as an XF-15 and a P-61A as an XF-15A, these being tandem two place 16 ton gross weight airplanes with, except for the new camera nose on the fuselage, the airframe of the basic P-61 fighter and the turbosupercharged Pratt and Whitney Double Wasp powerplants of the P-61C. Thirty-six production F-15A Reporters were procured in 1946, and these had a maximum speed of 440 mph at high altitude. With the advent of new jet reconnaissance designs the F-15As did not last long in the USAF.

PURSUIT/FIGHTER

In World War I American fighter pilots, then called pursuit pilots, flew foreign, single seat aircraft such as French Nieuports and Spads, along with a few British SE-5As and Sopwith Camels, in combat. These 1918 aircraft had one or two fixed forward firing rifle caliber machine guns, and together had an average high speed of 124 mph. The Spad, used also for some time post-war, had two guns and a high speed of 135 mph on 235 horsepower.

In the early 1920s Eberhardt turned out 50 British design Sopwith SE-5s for the Army, and Curtiss and Boeing learned something about pursuits in producing the designs of other companies by low bidding on open production contracts. Curtiss won the bid in 1920 on the Ordnance Engineering Company (Orenco) Model D and produced 50 planes; Boeing in 1921 won a Thomas Morse MB-3A pursuit production award for 200 planes. The two companies were thus ready in 1923 to compete with their Curtiss PW-8 and Boeing PW-9 entries (PW meant Pursuit, Watercooled), both powered with the same engine and with similar performance. Boeing clearly won out in production, however, with 113 airplanes to Curtiss' 28. These two companies were the Army pursuit producers into the early 1930s. Curtiss P-1, P-2, P-3, P-5, and P-6 Hawk biplanes, differing mainly in powerplant installations and ending with the attractive P-6E of 1932, competed with the Boeing P-12 series. Most of the Hawks had V-12 liquid cooled engines; all the Boeings had air cooled Wasps. Again Boeing won out big, producing 366 P-12s in several versions versus a total of 172 Curtiss Hawks.

In 1930 the Army took an excursion into the two seat fighter biplane area, but ordered only 25 aircraft from the Berliner Joyce Company and was not happy with the airplane.

Boeing again beat out Curtiss in a 1932-3 pursuit competition, and their low wing fixed gear wire braced monoplane P-26 with a radial engine became the Army's standard first line single seat pursuit with 139 built. But after that Boeing bowed out of the pursuit business.

A year or so later, 50 two seat cantilever low wing monoplanes with retractable landing gear, the PB-2As, were purchased from Consolidated. These planes used a turbosupercharged inline liquid cooled Curtiss Conqueror engine.

It appeared the Army could not make up its mind either on number of seats or on engine type. Soon however they decided on single seat fighters, and another Army pursuit plane competition in 1935-36 for a P-26 replacement resulted in a win for Seversky, a new producer of pursuits, with their P-35, while Curtiss and others lost. But the P-35 order was for only 77 airplanes (60 additional P-35As were later drafted by the Army from a Swedish order), while in 1938 Curtiss obtained an Army order for 210 of their similar, though usually better-liked P-36, in addition to later producing hundreds of export Model 75A versions. The two gun radial engined P-35 and P-36 were the standard Army fighters until 1939 when the Curtiss P-40 with a liquid cooled Allison engine won an order for 524 pursuit airplanes, the largest pursuit order between wars.

From then on through World War II the production of five major Army fighter types was quite astounding. The P-40 in various models came off the lines from May 1940 to November 1944 and production totaled well over 13000 fighter aircraft. The Bell P-39 Airacobra was in production from early 1941 through August 1944, and over 9500 were produced. Lockheed twin-engined P-38 production was initiated in the spring of 1941 and did not stop until the end of the war in August 1945. A total of slightly over 10000 Lightnings came off the lines. North American P-51 Mustangs started production in the fall of 1941; Republic P-47 Thunderbolts started in the spring of 1942; and well over 15500 of each type were manufactured through World War II. The war records of all these fighters, for better or worse, are well known. Less well known but also mass produced, especially for the Russians, were the 3300-plus Bell P-63 Kingcobras, and in more minor production for World War II over 700 P-61 Black Widow night fighters from Northrop.

The P-38 was known for long range and versatility; the P-39 criticized for lack of range and altitude performance; the P-40 was thought of as a rugged middle altitude stop-gap fighter in production too long; the P-47 praised for both high altitude performance and ground attack capability; and the P-51 Mustang lauded as the solution to the bomber escort problem. The Bell P-63 was especially short on combat radius; the big twin engined P-61 night fighter got mixed reactions—a fine flying aircraft, but too low and slow for a late World War II fighter.

Other more minor wartime types included the 272 Republic P-43 Lancers of 1941-42, high altitude fighters produced between the P-35 and the P-47 with many of these converted to photo work; the 144 Vultee P-66s, drafted into the Army from export contracts and detailed to China; and over 100 Douglas P-70s, A-20 light attack bombers modified as a stop-gap night fighter until the P-61 Black Widows arrived.

During the war a Bell P-59 fighter had been developed with jet engines, the first so powered in the US. These aircraft were used to train pilots for the Lockheed P-80 Shooting Star starting delivery in April 1945, the first American operational jet fighter which just missed World War II action but was in plenty of time for Korean War fighting. The P-80 used a centrifugal flow jet engine based on British design.

Another post-war aircraft was the North American P-82 Twin Mustang escort fighter with service introduction in early 1946 and having the distinction of being the last Air Force piston engine fighter. It was used early in the Korean conflict, mainly as a night fighter.

Republic Aviation started delivery to operating units of the first U.S. axial flow jet engined fighter, the F-84 Thunderjet, in June of 1947. These straight wing aircraft were also used in Korea mainly as fighter-bombers.

The USAF acquired its first swept wing jet fighter, the F-86 Sabre, starting in February of 1949. The Sabre acquired fame in jet dogfights with Russian MIG-15s in Korean skies and went through many modifications to become standard equipment for many years.

Introduced in May and September of 1950 respectively, the single-engine Lockheed F-94 Starfire all-weather fighter, a two seat stretch of the P-80, and the twin engine two place Northrop F-89 Scorpion jet night fighter increased the Air Forces' anti-bomber capabilities using nose radar and heavy machine gun and cannon armament respectively. Both were straight wing fighters, though the developed F-94C of March 1953 had a swept horizontal tail, and this version also carried pods of unguided rockets in place of guns.

The year 1952 saw introduction in June of the F-89D Scorpion version which had no guns but introduced wingtip pods for carrying unguided rockets with which to destroy bombers, and in October the F-86D Sabre Dog also became operational with the USAF Air Defense Command. The F-86D carried unguided rockets in a retractable belly tray and used a semi-automatic ground and fire control system to guide it to a bomber target using a lead-collision course attack plan. Later in 1955 advanced models of the Scorpion had different armament; the F-89H carried six Falcon guided missiles in new wingtip pods, and the F-89J carried nuclear-tipped rockets along with Falcons.

In January of 1954 a new swept wing version of the F-84 fighter was introduced into service, the F-84F Thunderstreak and its reconnaissance brother the RF-84F Thunderflash. Both of these types had a long service career with the USAF and many were exported to friendly foreign countries under aid programs.

A major step forward in technology came with introduction of the North American F-100A Super Sabre which, after initial difficulties in development, came into the USAF fighter inventory in the spring of 1955. Shortly afterwards in that summer a tactical fighter-bomber version, the F-100C, was introduced, and later the improved F-100D went to Tactical Air Command to start a long life and see combat in Vietnam. The F-100 series were the first fighters to attain supersonic level flight capability.

The next "Century Series" fighters were the very fast McDonnell twin engined F-101A and F-101C Voodoos of the late 1950s used by Tactical air Command as fighter-bombers. Reconnaissance versions, the RF-101A and RF-101C, followed into TAC service and performed many war missions in Vietnam. In addition, a two seat interceptor version was employed by the Air Defense Command as the F-101B into the early 1960s. Many F-101 and RF-101 aircraft served through the 1970s with the Air National Guard.

Getting a new century series jet fighter into operational service often proved difficult as illustrated by Convair's F-102A delta wing interceptor fighter which had a first flight in October 1953 and started operational service only in the spring of 1956, normally carrying armament of six Falcon guided missiles in a belly compartment. There was no gun armament. The F-102As lasted through the 1960s, including National Guard service. In addition, a few two place TF-102 trainers were produced.

Kelly Johnson's thin straight wing F-104A interceptor with Mach 2 speed and armament of an M61 Gatling gun and two Sidewinder missiles at the wingtips came into the USAF ADC in early 1958, and the next year the F-104C fighter-bomber, using bombs and missiles, entered the inventory. More F-104 production took place in Germany and Japan.

The big heavy (25 ton) Republic F-105 Thunderchief was the new USAF Tactical Air Command fighter-bomber with the fast-firing M61 gun and capable of carrying many combinations of bombs and rockets, both in an internal bomb bay and externally on store stations underwing. Entering service first in 1958, various "Thud" models were the TAC workhorses in the Vietnam War, and many were lost. There were both single and two seat versions.

A higher-performance replacement for the F-102A all-weather interceptor, the Convair delta wing F-106As coming into service in 1959 with automatic operation from takeoff to landing monitored by the single pilot, and capable of carrying nuclear and various homing missiles was the Air Defense Command standard interceptor against bombers for many years, with some still in service in the early 1980s having had many upgrades of equipment. Two seat versions (the TF-106As) were also produced.

Recognizing the U.S. Navy had a really good thing, in November 1963 the USAF put their version of the two place twin-engine McDonnell F-4 Phantom II into service as an F-4C (originally F-110) tactical fighter and RF-4C reconnaissance version. Probably as well known as any modern fighter, the 1400 mph Phantoms served the Air Force in several later versions and fought in the Vietnam war, initially having only missile armament aboard, but later equipped with a 20mm M61 Gatling gun. Phantoms shot down over 100 MIG aircraft in Vietnam skies, but many were lost, most often to ground fire. Production of F-4s continued into the mid-1970s with over 2500 put out, and they were used at one time in almost 50 squadrons. In addition, many Phantom variants were exported to friendly countries.

The small, relatively lightweight Northrop F-5 of 1964, costing much less than other U.S. fighter designs, was not bought in quantity by the USAF, but rather used to supply friendly countries with modern, easily maintained fighter forces, and hundreds were exported during the 1960s and 1970s in both single seat version and two seat fighter-trainer versions.

Although not a production aircraft (only four were built), the Lockheed YF-12A was so significant as to warrant inclusion herein. The ultimate interceptor aircraft with Mach 3 performance and missile armament, it was not produced because there were no enemy bombers to intercept, since the Russians had turned to ICBMs. First flown in 1962 and unveiled to the public in 1964, it was canceled three years later. The basic design, however, was used in the Lockheed SR-71 strategic reconnaissance aircraft.

The Convair F-111 tactical all-weather fighter was the swing-wing replacement for the F-105 and originally conceived as the

TFX in the early 1960s. Its development was substantially hindered by an initial requirement that commonality should exist between an Air Force version and a Navy F-111B version. The Navy versions were failures. The two place F-111A used an M61 20mm cannon, and could carry a heavy load of bombs, missiles, or extra fuel tanks. It was equipped with terrain-following and attack radar, along with an inertial bombing and navigational system. Operational in 1967, the F-111s were used for night attack missions in Vietnam. Late F-111 models were built in the 1970s, and several hundred went to the USAF. The F-111 is currently scheduled to go out of service.

The F-15 Eagle from McDonnell Douglas is the current USAF air superiority fighter. The F-15 is a 22 ton Mach 2.5 twin-engine single place aircraft with a 45 degree fixed swept wing, and is armed with a 20mm M61 six barrel cannon and up to eight missiles. First deliveries of operational aircraft took place in 1976, and by November 1989 the last of the 894 F-15s in several variants had been delivered.

The other current first line USAF fighter is the General Dynamics F-16 Fighting Falcon, a single place single-engine aircraft with a normal gross weight of about 14 tons and a high speed of over Mach 2.5. It has a single 20mm M61 Gatling gun and carries a Sidewinder missile on each wingtip. The first F-16A delivery to the USAF Tactical Air Command was in August of 1978, and a total of over 2200 have been delivered. Counting exports, over 3500 F-16s have been produced.

The F-117A "Stealth Fighter" is currently in service, however only a few dozen of this model have been built, and the final aircraft delivery was made in July of 1990. Built in the Lockheed "Skunk Works", the F-117A was designed to have stealth characteristics achieved by configuring the aircraft and using construction materials for minimum radar cross-section. The F-117A participated successfully in the initial U.S. attack on Iraq in the recent Gulf War.

After a competitive fly-off with the Northrop YF-23, the Lockheed Martin F-22A was selected for production as the latest U.S. air superiority fighter. It is a low radar cross-section single seat twin engine-aircraft, and a requirement for several hundred exists.

226.SPAD VII A rebuilt example of the famous SPAD pursuit of World War I, a design of French origin. A total of 1124 SPADs, mainly the VII and XIII variants, were purchased by the U.S. Army from the French to be used by the American Expeditionary Force. The early model VII, of which 189 were obtained starting in December of 1917, was soon supplanted, starting in early 1918, by the more numerous Model XIII. The SPAD VII had a single machine gun, weighed 1550 pounds, and was powered by a Hispano engine of 180 horsepower, providing a top speed of 127 mph. The more powerful SPAD XIII weighed slightly over a ton and had a "Hisso" engine uprated to 235 horsepower and a 135 mph top speed along with two guns. The 893 SPAD XIIIs became the standard pursuits, and many served post-war into the early 1920s with modifications.

227.THOMAS MORSE S-4C A restored example of the Thomas Morse S-4 pursuit trainer of 1917-18. After 100 S-4Bs had been built Thomas Morse produced a total of 497 S-4C models. Equipped with a single machine gun, the S-4C was a popular trainer for pursuit pilots. The S-4C weighed just under 1400 pounds and was powered initially by a 100 horsepower Gnome engine which gave a high speed of 100 mph. The Gnome was unreliable, so a change was made on the last 447 aircraft to an 80 horsepower Le Rhone powerplant, reducing speed to about 90 mph. A large number of S-4s on contract were canceled at war's end.

228.STANDARD E-1 Another pursuit trainer of 1918 was the Standard E-1 of which 128 were procured in 1918 by the Army. A compact biplane with a wingspan of only 24 feet, the 1140 pound gross weight E-1 was, like the Thomas Morse S-4, originally powered with the 100 horsepower Gnome engine, but after the first 30 had been delivered troubles with the Gnome powerplant dictated a change on the last 98 to a Le Rhone engine of 80 horsepower. High speed of the E-1 was about 100 mph. Postwar fighter squadrons used SPADs and SE-5As for first line fighters, however. Some E-1s were dubbed M-defence.

229.THOMAS MORSE/ BOEING MB-3A The 1919 Army home-grown fighter replacement for the SPAD, 54 of which—including four prototypes—were built by Thomas Morse as the MB-3, was successful enough to merit a further order for the compact little two-gun fighter. The big fighter order of 1921 for an improved MB-3 as shown in the photo was up for bids, and Boeing won, producing 200 of the MB-3A fighters from local spruce lumber for $6,617 per airplane. The last 50 planes had revised tail surfaces. The order made Boeing into a fighter manufacturer for many years. The MB-3A had a 26 foot wingspan, a 2500 pound gross weight, and was equipped with a 300 horsepower Wright Hisso engine which provided maximum and cruising speeds of 140 and 125 mph respectively. Service ceiling of the little fighter was 19500 feet.

230.CURTISS-ORENCO D Orenco meant the Ordnance Engineering Company of Long Island, which built four prototype wooden Model Ds with delivery starting in January 1919, with 300 horsepower Hisso H engines and a 147 mph speed. The Army decided to order 50 more Orenco D pursuits and put a contract out for bids. Curtiss won the bidding, and in 1920-21 manufactured the double bay biplane in production. The resulting pursuit plane had a gross weight of 2400 to 2800 pounds, a span of 33 feet, and a Wright Hisso engine of 300 horsepower powering the plane to a high speed of about 140 mph.

231.CURTISS PW-8 With Orenco D production behind them, Curtiss put out three prototypes of a new fighter with two bay biplane wings and a Curtiss watercooled D-12 engine with skin cooling on its wings. The photo shows the first prototype with pilot Russell Maughan leaning on the propeller and other friends. The influence of the Curtiss racers of the day was evident. With a 32 foot wingspan and a gross weight of about 2800 pounds, the PW-8 prototype shown here had a 440 horsepower engine and a maximum speed of 168 mph. This performance led to a contract for 25 production PW-8s (PW indicated Pursuit, Watercooled). The PW-8 production airplanes were heavier at 3150 pounds, had an uprated D-12 engine of 460 horsepower, and a revised fuel system.

232.BOEING PW-9C Based on experience in manufacturing the MB-3A pursuit, and taking advantage of newly developed welding techniques, Boeing developed a new Army pursuit called the PW-9 (ninth pursuit with watercooled engine) which had a first flight in June 1923. The PW-9 fuselage structure was of welded steel tubing instead of wood. Altogether 123 PW-9 pursuits were delivered, ending with the PW-9D of 1928. The PW-9C shown in the photo had a wingspan of 32 feet and a gross weight of 3170 pounds. Like all PW-9s, the C version was powered by a Curtiss D-12 engine of 440 horsepower, giving a top speed of about 160 mph and a service ceiling of 21000 feet. Two machine guns were fitted and two 122 pound bombs could be carried.

233.CURTISS P-1B The P-1 pursuit series, named Hawks by Curtiss, started in 1925 following development work with an XPW-8B airplane. Major changes from the PW-8 included new single bay tapered wings and a water radiator under the engine instead of skin cooling. A total of 150 Hawks were procured between 1925 and 1929, including 15 P-1s, five of which were converted to P-2s with new powerplants; 83 P-1A, B and C models with minor differences; and 52 AT-4 and AT-5 Hawk advanced trainers converted back to P-1D, E, and F pursuits by engine changes. The brand new P-1B shown in the November 1928 photo was powered, like the other Hawks, with a Curtiss D-12 engine of 435 horsepower, giving a high speed of 157 mph. The Hawks had a wingspan of 31 feet seven inches and weighed between 2850 and 2950 pounds loaded.

234.CURTISS P-3A The P-3 Hawks had radial engines. After an XP-3 and XP-3A powerplant test bed airplane was converted from a P-1A in 1926, first with an experimental Curtiss radial engine and later with a Pratt and Whitney Wasp, five P-3A Hawks were manufactured in 1928 using the 450 horsepower Wasp engine. The P-3As had a gross weight of 2790 pounds and a maximum speed of 153 mph. The photo shows one of the five P-3A pursuits fresh from the Curtiss factory with an external fuel tank under the fuselage adding to internal capacity. Later, ring cowlings were added around the cylinders of the Wasp engine.

235.CURTISS P-6 Another series of Curtiss Hawks were those using the new Curtiss V-1570 liquid cooled 600 horsepower engine, later known as the Conqueror. After two racer conversions used the engine in 1927, 18 P-6 hawks were produced by Curtiss in 1929 to 1930 with eight of these being revised to a P-6A version with a Prestone cooler instead of water. The airplane shown in the October 1929 photo has been so converted. The new engine installation required some refairing of Hawk fuselage lines. The P-6 airplane weighed 3150 pounds and used a V-1570-17 Conqueror of 600 horsepower to attain a high speed of 180 mph. These aircraft were soon to be revised again in service.

236.CURTISS P-6D A total of 13 P-6 type Hawks, one of them an experimental XP-6D, were converted in service with installation of turbosupercharged V-1570 engines of 600 horsepower to provide the Hawk with good high altitude performance, this done in 1931-2. These aircraft were service tested at the east coast Langley Field base. The photo shows one of the P-6D pursuits in May of 1935; the turbo unit is mounted along the side of the engine with the exhaust manifold feeding into the turbine section and the compressor exit air feeding into the carburetor via an intercooler also alongside. The P-6Ds weighed 3200 pounds loaded and had a maximum speed of 180 mph at altitude.

237.CURTISS P-6E After considerable experimentation with various prototypes Curtiss was able to obtain a July 1931 pursuit plane contract for 46 Y1P-22 models, which were shortly redesignated as P-6Es. Delivery of the P-6Es took place in 1932. A sleek biplane having a new powerplant arrangement with a small Prestone radiator, fixed guns buried along the fuselage sides, and single strut landing gear with panted wheels, the P-6E, shown running up in the photo, was a popular pursuit aircraft with the public. Using a 600 horsepower Curtiss Conqueror engine, the approximately 3400 pound gross weight aircraft had a maximum speed just under 200 mph and cruised at about 170 mph. Three of the P-6Es were modified for testing with powerplant, armament, and cockpit changes, these being labeled XP-6G, XP-6H, and XP-6F respectively.

238.BOEING P-12 The P-12 series was a particularly famous line of Army pursuit airplanes that originated in 1928 as a Boeing private venture with two prototypes going to the Navy. The Army was also interested on the basis of Navy testing, and in November 1928 Boeing received a contract for 10 P-12 pursuits, one of which was modified to an XP-12A to test certain features. The photo shows a Boeing Model 100 rebuilt to resemble the second Army P-12, Serial 29-354. The first flight of a P-12 occurred in the spring of 1929, and deliveries of P-12s took place that year.

239.BOEING P-12 P-12 pursuit planes were the mounts of several future USAAF commanders leading the air efforts of World War II, as is evidenced in this early 1930s photo of Major (later General) Carl Spaatz revving up his P-12 at Fort Sill, Oklahoma. Note the fairings behind each cylinder of the radial engine, which was a 450 horsepower Pratt and Whitney Wasp, giving the 2540 pound P-12 a maximum speed of 171 mph and a cruise speed of about 135 mph. Armament consisted of two fixed forward firing .30 caliber machine guns.

240.BOEING P-12D Continuous improvements were made in the P-12 Army pursuit. After 90 P-12B models were produced in 1930 with a revised landing gear and tail assembly and 96 P-12C Boeings were delivered the following year with a ring cowl around the engine, 35 P-12D versions like that in the photo, using a new higher compression Wasp engine, were produced in 1932. The D version weighed 2650 pounds, and its 525 horsepower Pratt and Whitney Wasp gave a high speed of 188 mph.

241.BOEING P-12E One of the most well known and popular of the old biplane Army pursuits was the P-12E, of which 110 articles were delivered by the spring of 1932 and utilized in first-line Army squadrons until supplanted by monoplane Boeing P-26s in the mid-1930s. The P-12E order was the largest since Boeing had obtained the contract for 200 MB-3As back in 1921. At the same time the U.S. Navy had been ordering the F4B- sea-going equivalent. The P-12E weighed about 2685 pounds fully loaded, used the same engine as the D, and had a similar maximum speed. A significant technical advance was the metal monocoque fuselage used in the P-12E.

242.BOEING P-12F The final production model of the P-12 series is shown in this run-up photo of a P-12F. Actually ordered with the P-12Es in March 1931, the 25 P-12Fs had a new F-series Pratt and Whitney Wasp rated at 600 horsepower, providing a high speed of 194 mph and better altitude performance than earlier models. Several P-12 airplanes were used for testing new features of powerplants, such as fuel injection and supercharging. Counting the USN versions and exports, this series of Boeing fighters provided a total production run of 586 airplanes, a record not to be equaled until World War II.

243.BERLINER-JOYCE P-16 An Army excursion into the two seat fighter area, and the only Army aircraft type ordered from Berliner Joyce, the P-16 was the subject of a 25 plane order after testing of a single prototype XP-16 in 1930. Production was completed in 1932. The planes were later redesignated as PB-1s for Pursuit, Biplace number 1. An unusual aspect of the P-16 was the upper wing being faired directly into the fuselage. The two place aircraft was not successful, and for a time the Army abandoned the idea, but it was destined to reappear. The P-16 weighed 4200 pounds and was powered by a Curtiss V-1570 Conqueror engine of 600 horsepower, giving the plane a top speed of about 170 mph.

244.BOEING P-26A Boeing became committed in the early 1930s to the all metal monoplane, such as its civil Monomail and military B-9 bomber. Between January and March of 1932 they built three prototype metal monoplane fighters later bought by the Army as service test Y1P-26s (then just P-26). In January 1933 the Army gave Boeing a production order for 111 P-26A pursuits to re-equip P-12 service squadrons. The P-26A was a mix of new and old features. Construction was all metal and the low wing monoplane configuration was employed, but the wing was wire-braced, the landing gear was fixed though faired in, and the cockpit was still open. Shortly wing flaps were added to cut down landing speed.

245.BOEING P-26A The photo depicts a restored P-26A at the Air Force Museum, and illustrates clearly the wire bracing between wing, fuselage, and landing gear. The ring cowling went around a 600 horsepower Pratt and Whitney Wasp radial engine. The landing gear was sprung independently of the wire brace structure in the gear fairing. The pilot's headrest was extended to provide for a turnover structure. The P-26A, sometimes called the "Peashooter", had a wingspan of only 28 feet, a gross weight of just about 3000 pounds, and its Wasp engine and two bladed fixed pitch propeller provided a top speed of 234 mph. It was armed with two .30 caliber machine guns and could carry up to 200 pounds of bombs under the fuselage.

246.Boeing P-26C A striking photo of a service P-26C "Peashooter" warming up at Boeing shows the under-fuselage bomb rack between the landing gear legs. A second production run of 25 aircraft came out in 1934-5, with two of these as P-26Bs with fuel injected Wasps and the others as P-26C versions with modified controls; these were later converted to P-26Bs. Eleven export versions of the P-26 type went to China and one got to Spain. A group of P-26As were turned over to the Philippine Army in late 1941, and these saw some of the futile action against the invading Japanese forces.

247.CONSOLIDATED PB-2A The unique factor about the Consolidated PB-2A was that it had two seats. It carried a pilot and a rear gunner. A descendant of the YP-24 and YP-25 low wing monoplane pursuit prototypes of 1931-33, the PB-2A started off as a P-30A ordered in 1934. After four P-30s (later PB-2s) were procured for service tests, a production order for 50 was obtained by Consolidated in December 1934, with manufacture taking place in the company's new plant in San Diego. All were delivered by mid-1936. The PB-2As were considered to perform well, but the rear gunner was ineffective in fighter maneuvers. These aircraft operated with the pursuit squadrons based at Selfridge Field in Michigan and Langley Field, Virginia.

248.CONSOLIDATED PB-2A The PB-2As had all the features of a then-modern pursuit type—low cantilever wing, all metal construction, retractable landing gear, and controllable pitch propeller—except it was a two-place aircraft. Armament consisted of two fixed forward guns and one rear flexible gun. With a wingspan of 44 feet and a gross weight of 5640 pounds, the PB-2A was powered by a turbosupercharged Curtiss V-1570 Conqueror engine rated at 700 horsepower at medium altitude, and had a high speed of 214 mph at sea level and 274 mph at 25000 feet, thus showing the effect of the turbosupercharging.

249.SEVERSKY P-35 After being rejected for a contract in 1935-6, Seversky entered a revised AP-1 pursuit design in 1937 and won a production award for 77 aircraft, the last plane being converted into an experimental XP-41. The 76 production airplanes were in the fully then-modern configuration of a single seat all metal low wing monoplane with enclosed cockpit and retractable landing gear. Actually, the gear was semi-retractable, since the wheels were housed in under-wing fairings when retracted. Powered with a 950 horsepower Pratt and Whitney R-1830 Twin Wasp engine and using a long chord NACA cowling, the P-35s in mid-1938 went to the Air Corps First Pursuit Group at Selfridge Field, Michigan.

250.SEVERSKY P-35 Flight view of the P-35 shows the characteristic elliptical wing planform and the two machine gun barrel blast tubes in the upper cowling. Small fragmentation bombs could be carried under the wings. The P-35 served along with the P-36 to fill out pre-World War II meager pursuit squadrons of the Army Air Corps. The Twin Wasp engine gave it a maximum speed of 281 mph. Gross weight was 5600 pounds and wingspan was 36 feet. The P-35A, a more powerful export version, was requisitioned from a Swedish order by the USA Army in 1941. Most all of the 60 P-35As went to the Philippines just before the Japanese invasion, and like other U.S. forces were shortly overwhelmed.

251.SEVERSKY P-35 The elliptical wing planform of the P-35, well shown in this August 1941 photo of the pursuit plane, was one way to achieve efficient aerodynamic span loading of the wing. Landing gear wheels were not fully covered when retracted. The P-35s all had variable pitch constant speed Hamilton propellers. The fuselage was fat enough that an extra man could be carried inside back of and below the pilot. The P-35 took no part in World War II other than being used as a fighter trainer. With the advent of that war pursuit planes were called fighters.

252.CURTISS P-36A The Army Air Corps companion fighter to the P-35 was the Curtiss P-36, of similar modern configuration and powered by the same engine type, though a more powerful version. After a prolonged pursuit competition among many contractors, several modifications of the Curtiss entry, and an initial loss to the Seversky P-35, Curtiss obtained contracts first for three service test Y1P-36s delivered in early 1937, and then that year a production contract for 210 P-36A models. The photo shows a restored P-36A in pre-war temporary camouflage now in possession of the U.S. Air Force Museum. The landing gear twisted 90 degrees so the wheel could lay flush in a wing pocket. Curtiss also produced hundreds of Model 75A export versions, many going to France to fight the German forces.

253.CURTISS P-36A Curtiss P-36A pursuits were considered to be very nice flying machines with an excellent rate of roll. Col. Charles Lindbergh flew both the P-36 and the P-35 and distinctly preferred the Curtiss airplane. The P-36A was the first of the Army's true 300 mph pursuits, achieving from 300 to 313 mph at a 10000 foot altitude with its 1050 horsepower Pratt and Whitney Twin Wasp engine and Curtiss Electric constant speed propeller. The first 177 aircraft of the 210 plane order were two gun P-36A models. Some were used for testing. The P-36A batch provided airframes for an XP-42 and an XP-40, as well as undergoing powerplant test variations.

254.CURTISS P-36C Thirty of the 210-plane pursuit order for the Army were delivered as the P-36C version of 1938-39, the main difference apparent from the A being the addition of two .30 caliber wing machine guns with external cartridge containers under the wings, making the P-36C a four gun fighter—still not impressive armament by World War II standards. The P-36C airplane shown was used for tests of camouflage paint schemes.

255.LOCKHEED P-38F The photo illustrates the twin engine, twin boom configuration of the famous P-38 Lightning. With long turbosupercharged engines, cooling systems, and main landing gear bays in the nacelles, it was easy to extend them aft a few feet and get to a twin boom arrangement. Turbosuperchargers and coolant radiators, as well as the nose guns, show up well. The P-38F of 1942 was the first version capable of carrying external stores underwing and also the first with a maneuvering flap. After 433 earlier models were produced, Lockheed built 527 F models. Powered by two 1325 horsepower Allison V-1710s, the nine ton gross weight P-38F had a maximum speed of just under 400 mph at altitude.

256.LOCKHEED P-38G The long, slim, high aspect ratio wing of the P-38 Lightning aided both climb and range performance. The P-38 was the long range fighter of the earlier part of World War II. The photo illustrates one of the 1082 P-38G Lightnings introduced in mid-1942 which employed new versions of the General Electric turbosuperchargers, uprated store station load capacity, and subsystem changes. P-38Gs fought in both North Africa and the South Pacific. Intercoolers for the turbosuperchargers were in the wing outboard leading edges.

256-A.LOCKHEED P-38J Lockheed started delivering the P-38J Lightning in the fall of 1943 and produced 2970 of these fighters. The photo shows the new, deep-chin look of the engine nacelles; additional internal fuel tanks were placed in outer wing leading edges with intercoolers now located in a new design in the engine nacelles. The photo was taken from the near P-38; the turbosupercharger in the left nacelle can be seen. Later J models incorporated underwing dive recovery flaps. The Allison engines were rated at 1425 horsepower each, wingspan was 52 feet, and normal maximum gross weight was over 16000 pounds, though ferry mission takeoff weight was over 21000 pounds. High speed at altitude was over 410 mph.

257.LOCKHEED P-38L The final and most produced version of the Lightning was the P-38L, of which over 3900 were delivered starting in June 1944. The photo of this aircraft was taken at Newark, New Jersey, in April of 1945. As on all production P-38s, armament consisted of four .50 caliber machine guns and one 20mm cannon all closely grouped in the nose. The L models were equipped with uprated engines having many automatic powerplant control features and were able to carry rockets underwing. P-38s were relatively expensive and hard to maintain and disappeared quickly after the war, whereas P-47 and P-51 fighters stayed around much longer. Weights and performance of the P-38L were similar to those of the J version.

258.BELL P-39D Flight photo of the first production model of the Bell P-39D Airacobra shows its clean lines. Armament consisted of a 37mm cannon firing through the propeller hub, two .50 caliber fuselage nose machine guns, and four .30 caliber wing guns. An Allison V-1710 engine of 1150 horsepower incorporating a single stage mechanical supercharger was located mid-fuselage and drove the propeller and reduction gearbox through a ten foot extension shaft. The P-39D entered service in April 1941, and with the P-40 took the brunt of early Pacific fighting in the air against the Japanese. Short range and limited effective altitude capability dogged the P-39. In addition, it was labeled a tricky plane to fly, but with its heavy armament turned into a good ground attack aircraft.

259.Bell TP-39Q The photo shows a renovated postwar P-39Q fighter in 1976, and illustrates the tricycle landing gear arrangement and the location of the exhaust stacks behind the cockpit at the engine location. Of about 9600 Airacobras produced, over 4900 were P-39Qs, with this final model first delivery in May of 1943. Almost 5000 P-39s were shipped to Russia. The P-39Q was powered by an Allison engine of 1200 horsepower. High speed of the Q was 385 mph, one of the fastest Airacobras; some earlier models were 20 mph slower. The P-39 was the smallest of the mass-produced Army World War II fighters with a wingspan of only 34 feet. Gross weight of production models varied little, running from 8200 to 8500 pounds.

260.CURTISS P-40 The initial production version of the XP-40 prototype, the latter being a P-36A airframe with an Allison V-12 engine for test. The P-40 won a 1939 pursuit competition, and shortly Curtiss was given an order for 524 aircraft; 200 of these were P-40s with 1040 horsepower Allisons and a high speed of 357 mph at 15000 feet. No armor or leakproof tankage was provided; armament was two .50 caliber machine guns in the nose and two .30 caliber wing guns. A typical gross weight for a P-40 with maximum fuel aboard (all internal) was 7173 pounds.

261.CURTISS P-40C A P-40C with its external centerline tank aboard is refueled in the Curtiss yard. After 131 P-40Bs were produced Curtiss went on to put out 193 C versions. By this time many export Tomahawks had been manufactured also. The P-40C went into service in early 1941 and was armored, had self-sealing fuel tanks, a centerline drop tank, and four wing guns besides the nose .50s. It was heavier; a typical aircraft weighed slightly over 8000 pounds with all tanks, internal and external, filled up. High speed fell to 340 mph.

262.CURTISS H81A-2 An export version of the P-40B in November 1940, known as the Tomahawk and destined for Britain. Hundreds of Tomahawks were exported; many fought in North Africa. The blast tubes for the two .50 caliber fuselage guns show above the engine; the carburetor air intake is in between. The lower scoop provided air to coolant and oil cooling radiators; air exit flaps are open. A ring and bead gunsight is located just ahead of the armored windshield. Landing gear wheels and shock struts twisted 90 degrees upon retraction so wheels could fit into wing pockets.

263.CURTISS P-40E-1 The Curtiss photograph shows a P-40E-1 (Lend-Lease P-40E) at the factory in the summer of 1942 with a non-standard belly tank under test. The E version was the result of an extensive redesign of the P-40, and first appeared in August of 1941. The P-40E had a new fuselage, a new version of the Allison engine with revised nose reduction gearing, and as a result shorter landing gear. Armament was completely changed to six .50 caliber guns in the wings and none in the fuselage. The E had an 1150 horsepower Allison V-1710 giving a high speed of 354 mph, and with a full standard external fuel tank weighed about 8700 pounds. The E was the first truly mass-produced version; including Kittyhawk I models, over 2800 of these fighters were delivered by Curtiss.

264.CURTISS H87A A Curtiss H87A Kittyhawk for export to Britain poses in slow speed flight for the company photographer as evidenced by the open cockpit canopy and cooling exit air cowl flaps underneath. The Army P-40s and their export Allison-powered equivalents were strictly middle altitude fighters since the engines incorporated a single stage supercharger like the Bell Airacobra. At 30000 feet the engine developed less than 500 horsepower. P-40F and L versions had Packard Merlin engines with single stage two speed mechanical superchargers which gave somewhat better altitude performance.

265.CURTISS P-40N The final operational version of the P-40 series with the long fuselage started with the P-40F and the revised canopy unique to the later N models looked little like an early P-40. Well over 5000 N models were produced in several sub-variants. With the P-40F and on the P-40s were known officially in the U.S. as Warhawks, while Britain kept the Kittyhawk name. The photo shows a P-40N-35 flying over upper New York state in July 1944, four months before the shut-down of the P-40 production line. There were many complaints that the P-40 was produced too long.

266.REPUBLIC YP-43 The Republic P-43 Lancer was an interim production fighter in the Seversky/Republic family between the 1936-37 P-35 and the 1942 wartime P-47 Thunderbolt. After 13 service test YP-43s were delivered in 1940, one of which is shown in the photo, production orders were obtained by Republic for 259 P-43 and P-43A versions which filled the company lines until the P-47 was ready. The P-43 fixed the mold for the P-47, having a turbosupercharged radial engine providing excellent altitude capability, though armament was limited—two .50 caliber nose and two .30 caliber wing guns. P-43s filled in meager USAAF fighter squadron ranks just pre-war, and some were sent to China. Powered with a 1200 horsepower turbosupercharged Pratt and Whitney Twin Wasp, the 7800 pound P-43 had a maximum speed of about 355 mph.

267.REPUBLIC P-47D One of the premier World War II fighters, the P-47 Thunderbolt was a big heavy airplane with a lot of firepower. Designed as a high altitude fighter with an engine turbosupercharger, it was often successfully used as a low altitude tactical fighter-bomber. The Thunderbolt shown is a P-47D-10, one of the early D "razorback" models in Italy in 1944. Prior to the D version a single XP-47B was first flown in May 1941, and 773 B and C models were produced. The P-47Ds were by far the most numerous. With a typical gross weight of about seven tons, a wingspan of almost 41 feet, and a 2000 horsepower turbosupercharged Pratt and Whitney Double Wasp engine, the razorback P-47D had a maximum speed at high altitude of about 430 mph.

269.REPUBLIC P-47D A later variant of the P-47D with a bubble canopy and a much-needed increase in internal fuel capacity, which respectively improved pilot visibility and combat radius capability, the latter always a P-47 problem with that big engine up front. Shown also are the large underwing stores pylons, which were much criticized as being too big. Armament in all P-47s consisted of eight .50 caliber machine guns, four in each wing. The turbosupercharger was located aft under the capacious fuselage, a large portion of which was crammed with air ducts. The length of the landing gear was determined by propeller ground clearance; it telescoped slightly upon retraction to fit in the wing. Over 12600 P-47Ds razorback and bubble canopy were produced during World War II from 1942-44.

268.CURTISS P-47G Two Republic factories, one on Long Island, New York, and the other in Indiana produced practically all Thunderbolts. A few were produced by Curtiss in a poor effort starting in October 1942 and labeled P-47Gs, roughly equivalent to the early P-47B and P-47C models; Curtiss manufactured six aircraft in 1942, 271 in 1943, and 77 in 1944 before March of that year when the contract was canceled. The photo shows a P-47G at Buffalo, New York, in the Curtiss yard, June 1943. Early Thunderbolts were excellent diving aircraft, but relatively poor in climb. The narrow chord-bladed "toothpick" Curtiss propeller shown was replaced in later versions by Curtiss and Hamilton propellers with wider "paddle" blades. These, together with engine water injection, improved climb capability substantially.

270.REPUBLIC P-47M The photo shows one of the three prototypes for the P-47M Thunderbolt, a "hot-rod" fighter version put out at the end of 1944 to equip only one fighter group in Europe, the 56th "Wolfpack". With the uprated Double Wasp engine of 2100 horsepower (2800 horsepower COMBAT rated at 32000 feet) the 130 P-47Ms were the fastest U.S. World War II production fighters, with a maximum speed of over 460 mph at high altitude. One mission intended for the P-47M was to chase German V-1 "buzz bombs" and shoot them down.

271.REPUBLIC P-47N The final version of the Thunderbolt was the long range P-47N designed for use in the Pacific and entering service in the fall of 1944. A few P-47N groups did see Pacific action late in the war. The P-47N shown was still in service post-war with a "buzz" number on the fuselage and the old-style rudder stripes revived from pre-war times. The N had a new wing containing fuel and the uprated Double Wasp engine of the M, along with changes to engine controls, landing gear, and structure. It was almost as fast as the M version, but heavier, and when loaded full of internal and external fuel (1000 gallons or more) had a combat radius of 1000 miles.

272.NORTH AMERICAN P-51A An early version of the famous P-51 Mustang line, the P-51A (Mustang II for Britain) was one of the Allison V-1710 engined models. A total of 1083 fighter versions and 500 light dive bomber versions (see the A-36A) were put out by North American from mid-1941 through early 1943. The fighters were shared by Britain and the U.S., with initial orders all going to England as the Mustang I. This 1981 photo shows a post-war P-51A painted in British markings. The P-51A was manufactured in 1942 with 310 delivered. Powered by a 1200 horsepower Allison V-1710 engine, this low and medium altitude fighter had a maximum speed of about 390 mph and gross weight in the long range fighter mission was 9653 pounds. Armament consisted of four .50 caliber wing guns, and there was a bomb rack under each wing.

273.NORTH AMERICAN P-51C After testing two prototypes in the U.S. in late 1942 and several in England, a joint decision was made to install the Packard-built version of the British Rolls Royce Merlin with a two stage mechanical supercharger to provide the Mustang with excellent altitude performance. The result was the P-51B and P-51C aircraft (the latter essentially a B built in Dallas, Texas), a total of 3738 being built in 1943. Armament was the same as the B model. A 1380 horsepower Merlin engine powered the B and C airplanes to a maximum speed of about 435 mph at 30000. A typical gross weight was just under six tons.

274.NORTH AMERICAN P-51D Probably the most recognized World War II U.S. Army fighter is the P-51D, many of which are still flying as the mounts of rich hobbyists. Modifications from the B and C Mustangs included a new engine version, a new bubble canopy for much-needed pilot vision enhancement, six wing .50 caliber guns in a revised, more trouble-free installation, a dorsal fin (except for very early variants), and stronger wing bomb racks. The gross weight of the P-51D varied from 9600 to 12400 pounds depending on the mission.

275.NORTH AMERICAN P-51D This flight photo shows a post World War II P-51D Mustang with buzz numbers on the fuselage side. The P-51 was a very competent fighter and solved the problem of long range escort for U.S. bombers in both Europe and the Pacific. North American factories turned out 7956 P-51Ds in 1944-45 in an exceptional production effort. Powered by a 1490 horsepower version of the two stage supercharged Packard Merlin engine, the D version had a high speed like the B and C of about 435 mph.

276.NORTH AMERICAN P-51H The final production variant of the P-51 Mustang fighter, the P-51H started delivery from the factory in early 1945, but was just too late for significant Pacific combat. A development based on the results of the P-51 light fighter program (XP-51F, G, J), the P-51H, 555 of which were built before contract cancellation, had a V-1650-9 Packard Merlin rated at 1380 horsepower for takeoff, and with a new low altitude COMBAT rating of fully 1930 horsepower for a very short period. The fastest production P-51, the 11000 pound gross weight H had a high speed of approximately 485 mph at high altitude.

277.NORTH AMERICAN P-51H Flight photo of the P-51H Mustang being used as a test airplane shows the striking lines of the aircraft. The heavier P-51D used essentially the same airframe as the early lighter Allison-powered Mustangs and therefore was more limited in structural tolerance to G loading. There were thus a considerable number of incidents of D models breaking up in extreme maneuvers. The P-51H airframe was completely redesigned to withstand the G loads normally used for Army Air Force fighters. It served for several post-war years with Air Force Reserve and Air National Guard units.

279.BELL P-59B America's first jet-powered aircraft, the P-59 Airacomet was a notable step forward, though not really a fighter suitable for air combat in late World War II. After three XP-59A prototypes and 13 YP-59A service test aircraft were put out in late 1942 and early 1943, a total of 20 P-59A and 30 P-59B twin jet airplanes were produced by Bell in 1944 and 1945, these to be used as trainers in the then-new Lockheed P-80 Shooting Star program. A batch of 50 more P-59Bs were canceled in October 1943. A large (45 foot six inch span) single place aircraft with jet engines in the wing roots, the P-59 was slower and less maneuverable than the latest piston-engined fighters.

278.BELL P-59B This Airacomet view shows how low to the ground a jet airplane could sit with no propeller clearance requirement and thus save on landing gear length and weight. First flight of an XP-59A came on October 1, 1942. The two engines were produced by General Electric Co. based on the British Whittle centrifugal flow engine. Jet thrust started at less than 1600 pounds and through development increased to a rating of 2000 pounds each in the GE J31 engines of the P-59B. With this thrust the high speed of the aircraft at high altitude was about 410 mph. Gross weight was approximately 13500 pounds. Armament, never used in anger, consisted of a 37mm cannon (like earlier Bell piston-engined fighters) and three .50 caliber machine guns in the nose.

280.NORTHROP P-61C The first and only U.S. Army night fighter of World War II, the P-61 Black Widow was a large three place twin-engined, twin boom aircraft first flown in May 1942. After two XP-61 prototypes and 13 YP-61 service test models were put out, production of 200 P-61A versions commenced, with the first A aircraft delivered in October 1943. After 37 P-61As were produced it was found necessary to remove the upper power turret because it caused tail buffet. A total of 450 P-61B Widows followed in 1944 with a longer nose and more underwing stores capacity. Most had the upper turret back after the problem had been cured. At war's end deliveries were starting of 41 P-61C models as shown in the photo. These planes had turbosupercharged engines. The P-61 had radar in the nose, four 20mm cannon fixed in the belly, and, in most aircraft, an upper power turret with four .50 caliber guns. It made novel use of wing spoilers and had admirable flying characteristics.

281.NORTHROP P-61C Flight photo of the final production version of the Black Widow shows the general P-61 configuration. Engines in the most-produced B model were Pratt and Whitney R-2800 Double Wasps each of 2000 horsepower and providing a high speed of about 360 mph at medium altitude. Gross weight of the B was 29876 pounds for the basic fighter mission, 34900 pounds for a long range mission, and up to 36900 pounds for a bombing mission carrying three tons of bombs externally. With a wingspan of 66 feet the P-61 was as big as a medium bomber. Northrop built the turbosupercharged 2100 horsepower Double Wasp powered P-61C to blunt criticism that the P-61 was too low and slow for a night fighter, but it was too late for the war.

282.BELL P-63A A follow-on fighter to the Bell P-39 Airacobra, the P-63 Kingcobra was of generally similar configuration but with all new parts. After two XP-63 and one XP-63A prototypes of late 1942 to mid- 1943, the initial P-63A production type was delivered in October 1943, with 1825 total aircraft in nine sub-variants of the A following. Big changes from the P-39 were use of a two stage supercharged Allison V-1710 engine and a laminar flow wing, the former to correct the poor altitude capabilities of the earlier fighter. Internal fuel capacity was limited, however, so the combat radius capability of the Kingcobra was severely penalized. At the end of 1944 delivery of a P-63C variant with a new engine version and a ventral strake commenced with a total of another 1427 fighters delivered. Two thirds of the P-63 aircraft were sent to Russia. None were used by the USAF in operational combat groups.

283.BELL P-63A Flight view of a P-63A Kingcobra shows the clean lines of the fighter. Handling qualities were generally good, and maneuverability, especially in rolling quickly and turning, were said to be exceptional. The P-63A was powered by a 1325 horsepower (1820 horsepower with water injection) Allison V-1710 V-12 engine with two stage mechanical supercharging, giving a maximum speed of about 410 mph at medium altitudes. Wingspan was 38 feet four inches, fighter mission gross weight ranged from 8442 to 9650 pounds, and fighter-bomber gross weight was 10550 pounds. Armament consisted of a 37mm cannon in the nose (like the P-39), two .50 caliber nose cowl guns, and two .50 caliber wing guns in underslung pods. Three underwing store stations were available, with one on centerline under the fuselage.

284.BELL RP-63C Bell produced 100 RP-63A and 200 RP-63C aircraft; one of the latter is shown in the photo. These planes were used as actual manned targets for training bomber gunners. Special thick skinning was employed and all armament was removed. The RP-63s would make simulated firing runs at U.S. bombers in training; gunners used special frangible bullets. Some targets were rigged with a light that turned on if the target was hit; these were called "pinballs". More realistic gunner training could hardly be imagined.

285.VULTEE P-66 The P-66 Vanguards got in the USAAF by being drafted in late 1941. The original Swedish order for 144 aircraft in 1940 was blocked by the U.S. government, and the fighter was assigned to the United Kingdom. With the advent of U.S. entry into World War II the Vanguards were assigned to U.S. west coast defense. When they were no longer needed in the U.S. most all were shipped to China. The maximum gross weight of the P-66 was 7400 pounds, wingspan was 36 feet, and the 1200 horsepower Pratt and Whitney Twin Wasp engine powered the aircraft to a maximum speed of 340 mph. Vanguard armament consisted of two .50 caliber nose guns and four .30 caliber wing guns.

286.LOCKHEED P-80A Notable as America's initial operational first line jet powered fighter, the Lockheed P-80 Shooting Star was ordered as a prototype XP-80 in June 1943 and flew first in January 1944 with a British engine. Successful tests led to orders for two more XP-80A prototypes and 13 service test YP-80As which flew in late 1944. Deliveries of production P-80As began four months before the end of World War II, and the type did not see combat. After contract cutbacks Lockheed built 977 A models, more than 150 of these being FP-80 photo planes. The P-80A was powered by an Allison J33 centrifugal flow jet engine of British and U.S. General Electric Company design rated at 4000 pounds thrust. High speed was 558 mph down low. Normal gross weight was 11700 pounds; wingspan was 40 feet, and armament comprised six .50 caliber nose machine guns.

287.LOCKHEED P-80B The P-80B was a 1947 updated version of the A model with the very important addition of a pilot ejection seat, weaponry changes, and more J33 jet engine thrust. These aircraft came along at the end of the P-80A line. The photo shows one of 240 P-80Bs with 165 gallon wingtip-mounted drop tanks and a drag brake extended under the fuselage. Normal gross weight was just under six tons and high speed increased to 575 mph at low altitude. Maximum speed decreased as altitude increased because available jet thrust went down faster than air density reduced. At high altitudes it was just about 500 mph.

288.LOCKHEED F-80C The final model of the P-80 Shooting Star was the C version, in 1948 redesignated from P-80C to F-80C (from pursuit to fighter). The 670 F-80C models started deliveries from Lockheed in late 1948. Weight and performance were quite similar to the P-80B. This version of the Shooting Star had wing store stations for rockets or bombs. When the Korean War started in mid-1950 the F-80C went into action early against North Korean invaders. In 1948 a two seat trainer version was designed as a TF-80C, later changed to the popular T-33 jet trainer.

289.NORTH AMERICAN F-82B As the last piston engine and propeller equipped USAF fighter, the two place P-82 Twin Mustang had an unusual twin fuselage configuration based on P-51H fighters with a new center wing section and a horizontal tail in between. As shown in the photo of P-82B Betty Joe, one main landing gear leg was under each fuselage. First appearing as an XP-82 just before the end of World War II, a large order for P-82Bs was cut back to 20, with these delivered by early 1946. The aircraft pictured was famous for having flown non-stop from Hawaii to New York in early 1947. The six gun F-82B escort fighter had a wingspan of 51 feet three inches, a normal gross weight of 25000 pounds, and two 1380 horsepower Packard V-1650 Merlin engines provided a high speed of about 450 mph.

290.NORTH AMERICAN F-82E In 1948 another escort fighter version of the Twin Mustang came into the Air Force inventory, this a batch of 96 F-82Es as shown in the photo which illustrates clearly the novel configuration. In this year the P-82s became F-82s, along with other redesignated fighters. The new aircraft had Allison V-1710 engines of 1600 horsepower each and served as frontline escort fighters for about two years. Gross weight was just under 25000 pounds and maximum speed was 465 mph at medium altitudes. The propellers rotated in opposite directions.

291.NORTH AMERICAN F-82G Other Twin Mustang versions flying in 1948 were the F-82F, F-82G, and F-82H models equipped with radar and all-weather flying equipment. These planes replaced the Northrop P-61 night fighters. The lineup of Twin Mustangs in the photo shows the radome installation mounted ahead of the wing center section and the wing ports for six .50 caliber machine guns. The decals on the blades show that Aeroproducts propellers were used with the 1600 horsepower Allison engines. High speed was 460 mph and gross weight was up to 25900 pounds. Procurement of F-82F, G, and H aircraft totaled 150 aircraft. Many F-82 types served in the early stages of the Korean conflict.

292.REPUBLIC F-84B The straight wing F-84 Thunderjet was America's second operational jet fighter. The Republic aircraft differed from the Lockheed F-80 in using a nose inlet for its Allison (originally General Electric) J35 axial flow turbojet engine instead of the cheek inlets of the F-80's centrifugal flow powerplant. After three XP-84 and 15 YP-84 service test aircraft , the F-84B, two of which are shown in flight, became the first production version, with 226 airplanes delivered by Republic starting in mid-1947 and into 1948. The Thunderjet had six .50 caliber guns , four in the fuselage and two inboard in the wings. The B model J35 engine of 3750 pounds of thrust gave the plane a high speed of about 585 mph. Gross weight was normally 16500 pounds, with a maximum weight about 10 tons. After the F-84B, Republic produced 345 F-84C and D versions in 1948-49 with few changes.

293.REPUBLIC F-84E In 1949-51 a new version of the Thunderjet was in production with a more powerful engine and more fuel capacity, along with underwing pylons for external stores. The photo shows one of 843 F-84Es carrying rockets for ground attack on pylons, along with wingtip fuel tanks. The 5600 pound thrust Allison J35 engine would power a clean 15000 pound F-84E to about 610 mph at low altitude. The final F-84 straightwing production model was the F-84G of 1951, 3025 of which were built, with over 1900 of those going to NATO allies under mutual assistance programs. The G model could carry tactical atomic weapons and had aerial refueling capability.

294.REPUBLIC F-84F In mid-1950, as a result of design work on a swept wing version of the F-84E with a 5200 pound thrust J-35 engine called a YF-96A, Republic revised the design to use a Wright J65 version of the British Armstrong Siddely Sapphire jet engine. The aircraft was designated an F-84F Thunderstreak which first flew in late 1952. A total of 2711 Thunderstreaks were delivered with almost half going to NATO countries. The six gun F-84F "Hog" was equipped for the low altitude bombing system (LABS), had aerial refueling capability, and most often carried two very large drop tanks which barely cleared the ground as shown in the photo.

295.REPUBLIC F-84F The photo shows both side and plan views of the Hog, so named since it seemed to take forever to get off the ground. F-84Fs served both SAC and TAC and later the Air National Guard, and some lasted into the early 1960s. The Wright J65 engine provided from 7200 to 7800 pounds of thrust depending on the model, giving the Thunderstreak a high speed of about 690 mph low down. The wingspan was 33 feet seven inches, and normal gross weight was slightly over 25200 pounds. Late F-84Fs incorporated "all-flying" horizontal tail surfaces along with speed brakes. Aside from Republic production, General Motors put out almost 600 Thunderstreaks. France, Belgium, and Germany also operated F-84F aircraft.

296.REPUBLIC RF-84F In 1953 Republic designed a reconnaissance version of the swept wing F-84. In order to fit cameras in the nose designers had to revise the J65 air intake system into one with wing root intakes, and the fuselage was lengthened over four feet. The new version was labeled an RF-84F Thunderflash, produced starting in the 1954-55 time period and coming off the Republic lines until the end of 1957. Of the over 700 reconnaissance versions produced, nearly 400 went to NATO countries. Performance and weights were similar to the F-84F. A few RF-84F aircraft were modified into RF-84Ks for carriage under B-36 bombers.

297.NORTH AMERICAN F-86A As the first U.S. swept wing jet fighter, the North American F-86 Sabre series was indeed notable, and gained much publicity during aerial battles of the Korean War. Originally designed as a straight wing jet in 1945, the aircraft was revised with a 35 degree swept wing based on a review of captured German test data. After a 1947 first flight and successful testing of two XP-86 prototypes with J35 engines the USAF in late 1947 ordered 221 production F-86A Sabres using the General Electric J47 engine of 5200 pounds thrust. The F-86A had a high speed of about 675 mph at sea level, and it could exceeed Mach 1 in a dive. Armed with six .50 caliber machine guns alongside the nose intake duct, the eight ton gross weight Sabre became the USAF first line fighter in 1949, thus replacing the F-80 Shooting Star.

298.NORTH AMERICAN F-86D The so-called Sabre Dog was an all-weather bomber interceptor version intended to shoot down Russian aircraft with air-to-air rockets. The rockets were unguided 2.75 inch folding fin Mighty Mouse types controlled by a then-sophisticated fire control system to guide the F-86D in its attack using nose radar. First delivered in late 1952, the 153 aircraft ordered in 1950 were initially designated F-95As and later changed to F-86Ds. After prolonged development the D models went operational in 1953 with USAF Air Defense Command, and a total of 2504 aircraft were built, some going to other countries. The Sabre Dog lasted through the 1950s into the early 1960s.

299.NORTH AMERICAN F-86D The photo shows an F-86D with its belly tray lowered in rocket-firing position just aft of the nose landing gear door. During an air attack the tray, containing 24 rockets, snapped down, rockets were fired, and the tray quickly retracted again. The Sabre Dog radar is in the nose radome with the jet engine intake duct below. Late versions of the F-86D were powered by a General Electric J47 afterburner-equipped jet engine, providing 5500 pounds of thrust normally and over 2000 pounds of additional thrust in the afterburner mode. Span was 37 feet, one inch; normal and maximum gross weights were about 18000 pounds and 20000 pounds respectively, and maximum speed was approximately 700 mph at sea level and almost 100 mph slower at high altitude.

300.NORTH AMERICAN F-86E The initial version of an improved F-86 Sabre was the F-86E going into production in early 1951. The major changes were in the flight controls, with power-operated linked slab-type horizontal tail surfaces and artificial control feel providing greater overall effectiveness. North American produced over 330 E models and Canadair started Sabre production as well, putting out over 100 aircraft. Flight performance of the F-86E was generally similar to that of late A versions.

301.NORTH AMERICAN F-86F A major improvement in Sabre performance, particularly in maneuverability at high altitude, was achieved by a change in wing design extending the leading edge, using wing fences, and removing the earlier wing slat installation. The F model started delivery in 1952, and when production ended in 1956 over 2500 of the F-86F Sabres had been produced. The aircraft was particularly effective against the Russian-built MIG-15 it encountered in Korean War air battles. Powered by a General Electric J47 engine of almost 6000 pounds thrust, the nine ton normal gross weight Sabre had a high speed of just under 700 mph. Several variants were produced, including two place trainer versions. Many F-86 aircraft were handed off to friendly nations in the late 1950s.

302.NORTH AMERICAN F-86H The last production F-86 Sabre version was the F-86H fighter-bomber with many changes including a new engine version, new fuselage, greater wing and tail span, and revised landing gear. In the later H models a major change was replacement of the six .50 caliber machine guns with four 20mm cannon. First flown in mid-1953 at Columbus and starting delivery in the same year, the 473 F-86H aircraft were employed in first-line service in 1954 and lasted through the 1950s. Powered by an 8900 pound thrust General Electric J73 engine, the H version had a high speed of almost 700 mph at sea level. It weighed from 21900 to 24300 pounds loaded. Service ceiling was over 50000 feet.

303.NORTH AMERICAN F-86L This version of the Sabre was an upgrade of the remaining F-86D models to extend life of the type. In 1956 a total of 981 were modified to F-86L standards by adding two feet of span to the wing and again modifying the leading edge with slots. In addition, new avionics compatible with the SAGE system were installed. The F-86Ls served first with the Air Defense Command and later with the Air National Guard; the photo shows a lineup of California ANG aircraft.

304.NORTHROP JF-89A In 1946 the USAF ordered two XF-89 Scorpion prototype two place twin engine all-weather fighters, which were tested in 1948-9. The next year deliveries of 48 production F-89A types started, with most converted to B models in 1951-2. These aircraft had six 20mm cannon and radar in the nose and could also carry bombs or rockets underwing. Wingtip fuel tanks were incorporated. The straight wing 18 ton gross weight aircraft was powered by two Allison J35 jet engines under the fuselage equipped with afterburners rated normally at 5100 pounds thrust and 6800 pounds in afterburner each. Maximum speeds were 640 mph low down and 570 mph at 35000 feet. The photo shows a JF-89A Scorpion used for special tests. In 1952 164 F-89Cs were delivered with uprated J35 engines.

305.NORTHROP F-89D The new D model Scorpion was an all-rocket equipped interceptor fighter for the Air Defense Command, with deliveries starting mid-1952 and continuing until 682 aircraft had been completed. All the 2.75 inch air-to-air rockets were carried in wingtip pods, 52 rounds on each side. New radar and other avionics were provided, and a new version of the Allison J35 engines was installed with 5400 pounds of normal thrust and 7200 pounds each in afterburner, giving a high speed of 635 mph. The F-89D was a big, heavy fighter with a wingspan of almost 60 feet and a gross weight of over 21 tons.

306.NORTHROP F-89J The last two models of the two place Scorpion were the F-89H and F-89J. The first of 156 F-89Hs started service in early 1956 armed with six radar guided Falcon air-to-air missiles plus 21 2.75 inch unguided FFAR rockets in each wingtip nacelle. Engine power was comprised of two afterburning Allison J35s rated at 5600 pounds thrust each normally and 7400 pounds each in afterburner. The final F-89J version was an updated F-89D, 350 of which were revised to carry two Genie nuclear rockets underwing. In addition, other rocket types could be carried in pods as an alternate to wingtip fuel tanks. The photo shows an F-89J banking away in a company shot.

307.LOCKHEED F-94A Lockheed solved the Air Force's need for a single engine all-weather jet fighter in 1949 by producing a YF-94 Starfire prototype comprised of Lockheed F-80 parts combined with a new two seat fuselage somewhat similar to that of the T-33 jet trainer (earlier the TF-80C). Production F-94A Starfires followed in 1950, and these 110 A versions had a nose radome with four .50 caliber guns underneath and a radar operator in the rear seat. Jet power from a 4600 pound thrust Allison J33 engine (6000 pounds in afterburner) pushed the 13000 pound F-94A to a high speed of just over 600 mph at sea level, reducing to about 550 mph at high altitude, the speed reduction due to engine thrust falling off faster than air density reduced with altitude.

308.LOCKHEED F-94B A minor modification of the Starfire in 1951 with greater tip fuel tank capacity and some avionics changes came out in early 1952 in time to participate in a minor way in the Korean conflict. A total of 356 F-94B models were delivered, but few got to Korea. The aircraft with the shark nose in the photo is somewhat unusual in not having large tip tanks aboard.

309.LOCKHEED F-94C The final version of the Lockheed Starfire was a major update of the basic design with a new engine, wing, tail, and armament system. The 387 F-94C jet interceptors used from 1953 to 1959 had a Pratt and Whitney J48 engine providing 6350 and 8750 pounds of thrust in normal and afterburner modes respectively. High speed was over 630 mph at sea level; speed dropped with altitude increases. Service ceiling was about 50000 feet. Interestingly, this Starfire had a straight wing and a swept horizontal tail. All guns were dispensed with; 2.75 inch unguided rockets were installed around the radar nose and additional rocket pods could be installed at the middle of each wing leading edge. Tip fuel tanks were normally installed.

310.NORTH AMERICAN F-100A The North American Super Sabre had the distinction both of leading off the "Century Series" of USAF fighters and of being the first fighter able to go supersonic in level flight. This achievement was gained via use of a highly swept (45 degrees) wing, a very clean design with nose inlet, and a new high thrust afterburning jet engine. After early stability problems had been corrected, the F-100A got into service in 1955 with a total of 203 aircraft built. A speed brake was fitted under the fuselage. Using the then-new Pratt and Whitney J57 engine of 10000 pounds thrust and 14800 pounds in afterburner, the F-100A at 25000 pounds gross weight could clock 850 mph at 35000 feet altitude. Armament consisted of four 20mm cannon. Two wing store stations were designed to handle drop tanks or bombs up to 1000 pounds each.

311.NORTH AMERICAN F-100C The next production version of the Super Sabre starting in 1955 was the F-100C fighter-bomber with a total of 476 aircraft produced from two factories. An uprated J57 engine of 10200 pounds thrust and 16000 pounds in afterburner provided a high speed capability of over 920 mph at high altitude and a ceiling of just under 50000 feet. Probe and drogue aerial refueling capability was included, and wing hard points added for carrying rockets, iron bombs, or nuclear weapons using the Low Altitude Bombing System (LABS). The photo shows an F-100C of the Washington DC Air National Guard.

312.NORTH AMERICAN F-100D A D model Super Sabre ordered in 1955 is shown as employed by the USAF Tactical Air Command. The F-100D fighter-bomber had a long life, first going into service in 1956 with some lasting well into the 1970s as Air National Guard aircraft. A total of 1274 F-100Ds were produced. The D version had the same engine as the C, and a top speed of about 900 mph clean. Normal gross weight was just over 17 tons. Changes included area increases in wing and vertical tail, an autopilot, and an expanding array of underwing store station loads, including two very large 450 gallon fuel tanks, Bullpup air-to-ground missiles, four Sidewinder air-to-air missiles, or standard iron bombs. Capability for nuclear weapon delivery was also included. The F-100Ds were used extensively for ground attack missions in the Vietnam War during the 1960s.

313.NORTH AMERICAN F-100F The F-100F derived from an earlier effort to design a two seat trainer version of the Super Sabre, resulting in a two place TF-100C of 1956. A cockpit section for a crew of two in tandem resulted in a longer canopy and a three foot longer fuselage. In 1957 production started on the F-100F version, and two years later a total of 339 two place Super Sabres had been delivered by North American. The F -100F had the general capability of the F-100D, but carried two 20mm cannon instead of four. A few F-100F versions were employed in Vietnam.

314.MCDONNELL RF-101A The McDonnell F-101 Voodoo started out as a long range escort fighter for USAF Strategic Air Command bombers, but was quickly switched to a Tactical Air Command fighter-bomber. Seventy-seven F-101As were delivered by late 1957 and the type was put into service that year. An additional 35 RF-101A reconnaissance versions came out about the same time—one is shown in the photo. A few RF-101As were employed in Vietnam. Powered by two Pratt and Whitney J57s giving 10200 and 15000 pounds of thrust normally and in afterburner mode respectively, the single place F-101A had a high speed of just over 1000 mph at altitude. The aircraft could carry a large fuel load, thus giving a combat radius capability of over 1000 miles. In the early 1970s the remaining RF-101As were updated and supplied to the Air National Guard as RF-101Gs.

315.MCDONNELL F-101B This version of the F-101 Voodoo was configured as a two place interceptor fighter for Air Defense Command and first flew in 1957. McDonnell produced 480 of the B version, including 72 dual control trainers. The F-101B went into service in 1959 and could carry two Genie nuclear rockets externally. A rotary bomb bay door allowed internal stowage of bombs or rockets, along with dual under-fuselage 450 gallon drop tanks. Powered by two uprated J57 jet engines of 10700 pounds normal and 16900 pounds afterburner thrust each, the F-101B could achieve almost 1100 mph at 35000 feet altitude. The interceptor could get to its ceiling of about 50000 feet in under six minutes. Some F-101Bs went to Canada as CF-101Bs. When ADC was finished with the type it went to the Air National Guard in the early 1970s and served for several years.

316.MCDONNELL RF-101C The F-101C version of the Voodoo, of which 47 were manufactured in the late 1950s, was used as a tactical bomber with beefed up structure: the RF-101C was the tactical reconnaissance version. These aircraft were used in reconnaissance missions over Cuba during the missile crisis of the early 1960s. Characteristics were similar to those of the RF-101A. The photo shows the large dropable fuel tanks located under the fuselage. The aircraft was almost 70 feet long, but its 35 degree swept wing had a span of less than 40 feet. Stall speed was about 200 mph at sea level.

317.MCDONNELL RF-101H At the end of the 1960s the F-101 aircraft were handed off to the Air National Guard. This 1967 photo shows an RF-101H Voodoo of the Nevada ANG and illustrates the 70 foot length of the single place reconnaissance aircraft. They were the largest of the USAF single seaters. The H models were upgrades of C variants performed for ANG use, and characteristics were similar to the C. McDonnell RF-4 reconnaissance Phantoms were the frontline replacements for the RF-101 Voodoos in the late 1960s and early 1970s.

318.CONVAIR F-102A Convair's Delta Dagger was a delta-winged bomber interceptor initially ordered in 1952. Ten YF-102 prototypes badly missed performance goals in 1953 and 1954, and the F-102 went through a major redesign, the principal change being an area ruled fuselage. In mid-1955, after tests of four YF-102As, the redesigned F-102A had its initial flight and production was underway. Armed with six Falcon missiles in a fuselage bay, the F-102A had a Pratt and Whitney J57 jet engine of 10000 pounds normal and 17000 pounds afterburner thrust. The Delta Dagger had a top speed of about 800 mph at 35000 feet, and it took approximately 10 minutes to reach that altitude.

319.CONVAIR F-102A The F-102A became the standard Air Defense Command interceptor starting in mid-1956 with 875 aircraft delivered in the next two years. In the mid-1960s a major updating program provided nuclear rocket firing capability and a new fire control system. As the F-102A aged and was replaced in ADC by F-106A aircraft, the Delta Daggers were turned over to the Air National Guard squadrons. The photo shows an F-102A of the Hawaii ANG in 1974. The aircraft is equipped with two 230 gallon external fuel tanks.

320.CONVAIR TF-102A A training plane was needed for the F-102A, and in late 1955 an order for 111 two place side-by-side versions was placed with Convair. These wider fuselage Delta Daggers had full operational capability and armament provisions so they could be employed right along with the single seat aircraft, and were periodically modified just like the F-102As, including new vertical tails, wing changes, and armament and avionics updates. Performance capabilities were generally similar to the F-102A.

321.LOCKHEED F-104A The F-104 Starfighter was novel in having a small, very thin, straight wing in a world of swept wing jet fighters. Aimed at a high speed of Mach 2.0, the single engine single place Starfighter needed an engine inlet shock control system to drop inlet air to a speed manageable by the engine compressor, and an afterburner to boost thrust. After tests in 1954 with two J65 powered XF-104 prototypes and a few YF-104As with the new General Electric J79 engine, 153 F-104A production fighters, the first flown in 1956, were ordered from Lockheed. All sorts of speed and altitude records were set by the F-104As; however, true ADC operational introduction occurred in early 1958. In two years the F-104As were relegated to the Air National Guard squadrons. Armament of the Starfighter comprised a 20mm cannon and two wingtip Sidewinder missiles. With the J79 engine producing afterburner thrust (rated 14800 pounds maximum), the F-104A could reach over 1300 mph at altitude.

322.LOCKHEED JF-104B As with other high performance fighter aircraft, the F-104 Starfighter needed a trainer version, and Lockheed obtained an order for 26 tandem two place F-104B models with similar operational capability to the A. First flight was in February 1957, and all aircraft were delivered by the end of 1958. The JF-104B photo (the J meaning a special test aircraft) shows the spiked engine inlets, the canopy for two cockpits, and four external fuel tanks, two at the wingtips and two inboard. When the F-104As went out of U.S. service, the F-104Bs were exported to Taiwan.

323.LOCKHEED F-104C The C version of the Starfighter was designed as a fighter-bomber for the Tactical Air Command, and deliveries of the 77 F-104Cs started in late 1958. Important changes included a provision for probe-drogue refueling, a new weapons fire control system, and bomb-carrying (including nuclear) capability from the fuselage. The 20mm high rate of fire cannon and missile firing capability of the F-104A version were retained, and performance of the 23000 pound gross weight fighter-bomber was also similar to that of the F-104A. In the middle 1960s the F-104Cs started going to the Air National Guard, where some lasted almost ten years.

324.LOCKHEED F-104D The F-104D Starfighter was the tandem two seat trainer version of the F-104C. This September 1970 photo shows a D model with open cockpit canopy sections and emphasizes the small wing size. Wingspan was only 22 feet. Only 21 of these trainers were procured by the USAF. The wingtip fuel tanks carried 170 gallons each.

325.LOCKHEED F-104J The photo shows an example of a Japanese F-104J, one of many manufactured under license by Mitsubishi into the late 1960s. Though there was limited F-104 production for the USAF, many hundreds of these aircraft were manufactured under license in Europe for the air forces of West Germany, Belgium, and the Netherlands. Other countries getting F-104 versions included Canada, Italy, Norway, and Denmark. In Germany, F-104 crashes were quite frequent, and a furor was raised over the fact.

326.REPUBLIC F-105B The Republic F-105 Thunderchief was designed to replace the company's F-84F as a fighter-bomber for Tactical Air Command. In February 1954 15 F-105As were ordered with Pratt and Whitney J57 engines, and a YF-105A first flew in late 1955. A switch was made to the more powerful Pratt and Whitney J75 with 24500 pounds afterburner thrust for production F-105B aircraft, of which 75 were built in 1958-59. A large, heavy fighter with a length of 63 feet and a gross weight of 24 tons, the "Thud" could clock about 1400 mph at 37000 feet and over 850 mph at low altitude. A 20mm Gatling-type gun was mounted in the nose. The weapon load capability was phenomenal for a fighter at up to 12000 pounds. An internal bay could carry a nuclear weapon or fuel tanks, and external store stations could mount a great variety of weapons or tanks.

327.REPUBLIC F-105D With an integrated flight and fire control system, the next model of the Thunderchief, the F-105D, had a high degree of all-weather capability provided by the new avionics. A centerline bomb rack could be installed as an alternate to use of the internal weapons bay, thereby adding to the combinations of possible externally carried weapons. The initial F-105D was flown in 1959, and production of over 600 D aircraft continued through 1963. The F-105D "Thuds" were used extensively as tactical bombers in the Vietnam War and suffered substantial losses to anti-aircraft gun and missile fire. The uprated J75 engines produced 16000 pounds of normal thrust with 24500 pounds in afterburner, and F-105D high speed was similar to the B model, as were gross weights. Many F-105Ds went to Air National Guard squadrons in the 1970s.

328.REPUBLIC F-105F The F Thunderchief was a two place model with a longer fuselage and a taller vertical tail appearing in 1963. A total of 143 two seaters were produced, having weapons capability generally equivalent to the other Thunderchiefs. Many F models were converted as Wild Weasel aircraft and employed in Vietnam to locate and destroy enemy radar installations. The photo of this F-105F was taken in May 1964 prior to Wild Weasel conversion and shows the aircraft with centerline and wing-mounted external fuel tanks. The 20mm cannon port can be seen at the left side of the nose.

329.REPUBLIC F-105F A manufacturer's flight view of a "clean" F-105F Thunderchief taken in July 1963, less than a month after F model first flight, shows the low-set horizontal tail and the petal-type afterburner enclosure, which also opened to form a speed brake. Some of the F-105F aircraft were modified to F-105G models with revised avionics and missile weaponry, and these served through the 1970s.

330.CONVAIR F-106A The F-106A Delta Dart was a very extensively revised version of the earlier delta wing F-102A. The wing was slightly smaller, though similar in span; the fuselage was completely new, as were the vertical tail and landing gear. The F-106A employed the Pratt and Whitney J75 engine of 17000 pounds and 24500 pounds thrust without and with afterburner respectively, providing a high speed of well over 1300 mph at high altitude. Gross weight was 34500 pounds, wingspan was just over 38 feet, and the aircraft had a length of 71 feet. The ceiling was over 50000 feet. The Delta Dart was an all-weather interceptor capable of fully automatic flight under ground control, and could carry various air-to-air missiles, including those equipped with nuclear warheads. USAF Air Defense Command was equipped with these interceptor fighters, 277 of which were built, from 1960 well into the 1970s.

331.CONVAIR F-106B As with all the previous "century series" fighters, the F-106A Delta Dart came out in a two seat variant, with 63 B models delivered by Convair. The crew of two were seated in a tandem arrangement as opposed to the side-by-side seating of the earlier TF-102A two seater Delta Dagger. The F-106B models were used as trainers, but had full operational capability like the F-106A if required.

332.MCDONNELL F-4C The USAF in 1962 decided to procure a version of the F-4 (was F4H-1) Navy-developed Phantom II all-weather fighter. The first Air Force version was designated an F-110A, but was shortly redesignated an F-4C, with a first flight in May 1963. In 1964-66 the USAF obtained 583 F-4C Phantoms, and these were employed in Vietnam where dozens were lost to AA. The C model had no built-in guns, only missiles, so a gun pod had to be adapted. Powered with two General Electric J79 jet engines of 10900 pounds normal thrust each and 17000 pounds in afterburner mode, the two place dual control F-4C had a high speed varying from over 800 mph at sea level to over 1400 mph at 38000 feet, along with a service ceiling of well over 50000 feet. The Phantom was armed with four semi-recessed belly positioned Sparrow missiles, along with a fuselage centerline store station and four wing pylons for various external fuel tanks, bombs, or missiles, and could carry a nuclear store.

333.MCDONNELL RF-4C Along with the F-4C McDonnell delivered over 500 RF-4C reconnaissance versions of the Phantom from May 1964 to the early 1970s with photo, side-looking radar, and infra-red sensor capability instead of weaponry. These aircraft were to replace the RF-101s. Performance and weights of the recon version were similar to those of the F-4C. Wingspan was 38 feet five inches and length 63 feet. The aircraft in the photo was in the Nebraska Air National Guard in April of 1977, and carries external fuel tanks on outboard stores pylons. Several hundred F-4C and RF-4C Phantoms were still in the Air Force and Air National Guard units in the mid-to-late 1970s.

334.MCDONNELL F-4D The next Air Force Phantom version was the F-4D, first flown in December 1965, with 825 aircraft delivered in the mid-to-late 1960s. Changes included updating revisions in both avionics and armament. Weights and performance were little different from the F-4C. The D version also served in Vietnam in the late 1960s. Although Phantoms engaged in some fighter-versus-fighter air battles, most were employed in ground attack missions. The D model was still armed basically with missiles only, Sidewinders, Falcons, and Sparrows, but external gun pods could be adapted. F-4D aircraft were exported to several countries in the 1960s, including Iran and South Korea.

335.MCDONNELL F-4E The final basic USAF version of the F-4 Phantom series was the multi-role F-4E, used for air superiority, close support, and interdiction missions. The F-4E first appeared in October 1967. The type was used for a period by the USAF Thunderbirds Flight Demonstration Team. A significant change was inclusion of a high rate of fire M61 20mm cannon designed integrally with the aircraft. Other changes included new radar and other avionics, slats back in the wings, and uprated J79 engines, along with additional fuselage fuel cells. The F-4E also served in Vietnam late in the war. Over 800 E models reached USAF service with production continuing into the 1970s. A considerable number of F-4Es were revised to F-4Gs with changes in avionics, including electronic counter-measures (ECM) equipment. Large numbers were exported to friendly nations such as Israel, Turkey, and Australia. The General Electric J79 engines of the F-4E were rated at 11900 and 17900 pounds thrust in normal and afterburner modes respectively, providing a clean aircraft maximum speed of about 900 mph at low altitude and almost 1500 mph at high altitude.

336.NORTHROP N-156F The N-156F shown was one of three prototypes of the Northrop F-5A Freedom Fighter, initially a private venture, for which the USAF contracted in May 1958. The aircraft was a single seat fighter version of the Northrop two place T-38 talon trainer developed earlier. After a late 1962 USAF production contract for 142 aircraft, the first F-5A flight took place a year later. The F-5A was designed as a relatively inexpensive light fighter for small, friendly nations. Powered by two General Electric J85 afterburning jet engines of 2700 and 4100 pounds of normal and afterburning thrust respectively, the 20000 pound gross weight F-5A had a high speed of just over 900 mph at 36000 feet. Basic armament consisted of two 20mm M-39 cannon in the nose and two wingtip-mounted Sidewinder missiles. Five store stations, one on centerline and four underwing, could carry a variety of weapons. A total of 621 F-5As were delivered; many went to NATO countries, and some went to South Vietnam.

337.NORTHROP F-5B The F-5B was a two place fighter-trainer version of the F-5A which first flew in February 1964. By late 1971 Northrop had orders for 134 F-5B aircraft. When F-5A aircraft were supplied to a country, F-5B two seaters were also provided for training purposes. One F-5B was modified in 1969 to test features to be incorporated in the later F-5E fighter.

338.NORTHROP CF-5A The photo shows an example of an export version of the F-5A Tiger fighter, the CF-5A produced by Canada using higher thrust J85 engines along with inclusion of aerial refueling capability. Canadair manufactured 115 of these fighters from 1965 to 1970. The high speed of the Canadair version was slightly over 1000 mph, and maximum climb rate was 31000 feet per minute. No external stores are mounted on the aircraft shown, but the centerline stores pylon can be seen projecting beneath the fuselage belly. More than 1150 F-5 types have been manufactured.

339.NORTHROP F-5E The advanced version of the F-5 fighter was the F-5E Tiger II, which was selected in November 1970 as the International Fighter Aircraft to follow the F-5A. Changes included uprated J85 engines each of 5000 pounds afterburner thrust, maneuvering flaps, wing leading edge flaps, and a leading edge extension. First flight took place in August 1972. Armament was the same as the F-5A, including the two 20mm nose guns, two tip-mounted Sidewinders, and five multi-use store stations. Wing span was 26 feet, eight inches, and length was 48 feet, four inches. Maximum takeoff weight was just over 24000 pounds, and maximum level flight speed at 36000 feet hit Mach 1.6. Service ceiling was 54000 feet. Over 950 F-5Es were delivered to a variety of countries.

340.CONVAIR F-111A A controversial twin engine swing-wing tactical fighter with side-by-side seating for a crew of two, the F-111A flew first with wings extended to their 63 foot span in December 1964. A total of 18 development and 141 production aircraft were turned out and started service in 1967. The photo shows an F-111A in flight with wings extended in slow speed mode. Powered by two Pratt and Whitney TF30 afterburning turbofan engines of 21000 pounds maximum thrust each, the F-111A had a high speed of 1450 mph at over 50000 feet and 900 mph at sea level. Normal gross weight was 83000 pounds; overload takeoff weight added up to 45 to 49 tons. A 20mm M61 Gatling gun was located in a weapons bay; bombs or missiles could be carried in the bay as alternatives. Four wing pylons, the inboard two swiveling with wing swing, could carry fuel tanks or bombs. A sophisticated all-weather bombing and navigational system was employed, and equipment for low level terrain-following flight was included.

341.CONVAIR FB-111A The two place strategic bomber version of the F-111 for the Strategic Air Command, having a first flight in 1968 with a total of 76 aircraft delivered between late 1968 and mid-1970. The FB-111A carried two SRAM missiles in an internal weapons bay, along with iron or nuclear bombs or fuel tanks on six wing pylons. No guns were carried. Powered by two uprated Pratt and Whitney afterburning turbofans of over 20000 pounds of maximum thrust each, the high speed of the FB-111A was about 1450 mph at high altitude and over 800 mph at sea level. Airframe components were strengthened for a gross weight of over 110000 pounds. The aircraft pictured is from the 509th Bomb wing at Pease Air Force Base in New Hampshire, and was photographed in May of 1971.

342.CONVAIR F-111D The photo shows an F-111D in July 1980 from the 27th Tactical Fighter Wing at Cannon Air Force Base in New Mexico. The D model was similar to the F-111A, with upgrades in navigation and weapons delivery systems. The first airplane was delivered to the USAF in mid-1970. A total of 96 aircraft were completed, with final delivery in early 1973. This version was powered by TF30 turbofans of 19600 pounds maximum thrust each.

343.CONVAIR F-111F The final fighter-bomber version of the F-111 series, the F-111F was similar to the F-111D, but with revised avionics and advanced P-100 versions of the Pratt and Whitney TF30 turbofan engines with each producing a maximum thrust of 25000 pounds. A total of 106 F-111Fs were manufactured, and deliveries took place from 1972 until 1976. The F model could clock 1400 mph at 40000 feet and 900 mph at sea level, and climb rate was initially over 40000 feet per minute. A 20mm cannon or other weapon could be carried in the internal bay. Up to four pylons per wing could be mounted, the inboard pylon swiveling as the wing swung back, while the others were fixed and jettisonable. Fuel tanks, iron bombs, nuclear weapons, and special "smart" bombs or jammer pods could be carried on these pylons. A total of well over 500 F-111 type aircraft in at least six versions have been produced; some took part in the Gulf War. The type is programmed to be phased out of the inventory.

344.LOCKHEED YF-12A The experimental/service test interceptor version, first flown in 1962, of Lockheed's basic A-11 Mach 3.0 high altitude two place aircraft design, publicly displayed first in September 1964. The YF-12A program was discontinued in 1967 when the Russians turned from manned bombers to intercontinental ballistic missiles. Some YF-12A aircraft were tested with air-to-air missile armament. Powered by two Pratt and Whitney J58 afterburning turbojet engines of 32500 pounds thrust each, in May 1965 a YF-12A showed a high speed of 2190 mph at over 80000 feet. The photo shows the YF-12A configuration with a long slim fuselage, basic delta wing planform, twin engine nacelles with shock cone center bodies, variable area exit nozzles, and all-moveable sloped vertical tails.

345.LOCKHEED SR-71A The two place twin engine strategic reconnaissance version of the basic A-11 design, the SR-71A first flew in December 1964 with deliveries of operational aircraft beginning in January 1966. A variety of reconnaissance sensor packages could be carried. Using the same J58 afterburning turbojets as the YF-12A, the SR-71A, a 140000 to 170000 pound gross weight aircraft, had a high speed of over 2000 mph at 80000 feet. At least 21 SR-71s were produced. The SR-71B was a tandem two seat trainer version delivered in January 1966 and lost in an accident. It was later replaced with an SR-71C training variant. The SR-71 was a large aircraft, with a wing span of 55 feet seven inches and an overall length of 107 feet five inches.

346.MCDONNELL DOUGLAS F-15A Designed as an air superiority fighter, the basic single place twin engine F-15 Eagle flew first in July 1972 and went operational in early 1976. Since that time a total of 894 F-15A, B, C, and D versions have been delivered, the last of these in November 1989. Powered by two 23900 pound thrust Pratt and Whitney F100 afterburning turbofan engines, the F-15 is basically a Mach 2.5 aircraft in clean interceptor configuration carrying four AIM-7 missiles. Maximum gross weight of the F-15A is 56500 pounds; the C version can go to 68000 pounds. The F-15C arrived in early 1979 and the D in June 1979. The F-15C had new avionics and more internal fuel, along with external "conformal" fuel tanks. The F-15B and D are two seat versions of the A and C versions, respectively, with the second man being a weapons system operator. F-15s have been exported to such countries as Israel and Saudi Arabia.

347.MCDONNELL DOUGLAS F-15A The F-15 Eagle has a basic armament of one 20mm M61 cannon with 940 rounds of ammunition, fuselage flank ejectors for four AIM-7 Sparrows or AIM-120 AMRAAM, a centerline pylon rated for a 4500 pound load, and two outboard pylons under each wing to carry fuel or sensor pod loads. In October 1989 the F-15E Enhanced Eagle two-seater became operational, with 209 aircraft being delivered by mid-1993. The E model is a dual-role fighter/attack type. Powered by either Pratt and Whitney F100 or General Electric F110 turbofans (the USAF uses the F100), the F-15E has a maximum weapon load of 24500 pounds, and a maximum takeoff gross weight of 81000 pounds. High speed is Mach 2.5, and combat radius is 790 miles.

348.GENERAL DYNAMICS F-16A The F-16 Fighting Falcon was selected in January 1975 as the new USAF low cost, very maneuverable lightweight single seat fighter. With a single engine and a normal gross weight little more than half that of the F-15A, the F-16A is indeed a simpler aircraft. Powered by a Pratt and Whitney F100 turbofan of 24000 pounds thrust, the 23360 pound gross weight aircraft (35400 pounds overload) has a maximum speed of about 1350 mph at 40000 feet and 915 mph at sea level, and can climb to over 60000 feet. Wingspan is only 32 feet 10 inches (over missile fins), with length 47 feet eight inches. Basic armament is comprised of an M61 20mm cannon at the left side of the fuselage and two wingtip-mounted Sidewinder missiles. External underwing store stations (one on centerline and six others) can support drop tanks, Sidewinder missiles, or bombs. Production F-16As went into service in early 1979. A two place F-16B has the same weapons capability but less fuel.

349.GENERAL DYNAMICS F-16C An uprated version of the F-16 Fighting Falcon, the F-16C production brings the total of F-16s for the USAF to over 2200 fighters. Counting exports, over 3800 have been built. The C model is powered by either a Pratt and Whitney F100 turbofan at 29100 pounds of thrust, or a General Electric F110 turbofan of 29000 pounds thrust. Normal and maximum gross weights of this model are 27185 and 42300 pounds respectively. Performance is generally similar to the A version. The F-16D is a two seat variant of the F-16C. A large number of F-16s have been provided to friendly nations.

350.LOCKHEED F-117A The F-117A has been designed as a "stealth" attack-fighter with an exceptionally low radar cross-section. This objective has been accomplished by careful shaping, along with use of radar absorbing materials. First flight took place in June 1981, and initial USAF operation took place in the fall of 1983. There were five prototype aircraft and a production run of 59 articles, each costing $42.6 million, with deliveries ending in mid-1990. Powered by two 10800 pound thrust General Electric F404 turbofan engines, the F-117A achieves a high speed of 646 mph or Mach .90. Two 2000 pound bombs or missile alternates can be carried in a weapons bay. Normal loaded and maximum gross weights are 48000 and 52500 pounds respectively. The F-117A can carry 5000 pounds of weapons for a 690 mile mission radius. Wingspan is 43 feet four inches.

351.LOCKHEED MARTIN F-22A A new air superiority fighter with a production aircraft first flight slated for May 1997, the single seat F-22A is programmed for a requirement of 384 aircraft. Boeing has responsibility for the wings. Powered by two Pratt and Whitney F119 afterburning turbofan engines developed from the F100 turbofan, the F-22A has a maximum speed using afterburners of Mach 1.7 at 30000 feet and 920 mph at sea level. Wingspan is 44 feet six inches, and maximum takeoff gross weight 30 tons. A two seat F-22B version is also under development.

LIAISON

In early 1942 the Army L- (for liaison) designator was born, and all of the lighter unarmed observation (O-) airplane types were redesignated into the L- classification. The need for a type of light, agile, slow-flying aircraft capable of very short field operation in close cooperation with Army field units for local transport and field commander inspections, fire support direction, and many miscellaneous tasks was recognized in early 1939 when a competition was held for a light two place Army cooperation airplane. The winner was the Stinson O-49, later redesignated as the L-1 Vigilante, the first, biggest, and heaviest of the early production L- series aircraft. It was as big as the O-47 standard observation airplane then in service, but had one third of the power and half the gross weight, and importantly, much better short field performance for improved forward area operations. Many were later converted to ambulances, but then came the revolution! After testing some civilian two place Aeronca, Taylorcraft, and Piper lightplanes in 1941, the Army realized that was just what they wanted as liaison aircraft, and they were cheap, too. In 1942 and early 1943 they ordered and received no less than a total of just under 9000 Taylorcraft L-2, Aeronca L-3, and Piper L-4 "Grasshoppers" (well over 5600 Pipers alone). These were 1200 pound aircraft with 65 horsepower engines and 90 miles per hour high speeds. They performed a myriad of tasks throughout the war, and had a number of hair-raising escapes from enemy aircraft. In addition to the most-produced trio above, Interstate in 1943 produced 250 of their two place Cadets as Army L-6s.

The other mass-produced wartime liaison airplane, a thousand pounds heavier and almost three times as powerful, was the Stinson/Vultee L-5 Sentinal, a military development of the civil Model 105 Voyager with fuselage modifications. From 1942 until 1945 about 3300 Sentinals were produced, 83 going to the Navy. They were used extensively both in World War II and the Korean conflict; some were used as ambulances and photo planes.

After World War II updated versions of Grasshopper liaison planes were still important in Army operations, and in the late 1940s over 600 Aeronca L-16s, updates of the wartime L-3 and a development of the Aeronca Champion, were procured. Most went to the Air National Guard and were later used for Civil Air Patrol duties. At about the same time, over 200 North American/Ryan Navions were purchased, some for the Army and some for the USAF, and in the early 1950s over 700 more Pipers were produced for the Army as the L-21 with more engine power than on the L-4; these served into the 1960s.

By far the most produced (over 3000 aircraft) true liaison plane of the modern era was the Cessna L-19 Birddog, an Army version of the civil Model 170. The Birddog was the last of the true two place lightplanes and served the military forces of many nations along with those of the US.

Several other aircraft types were procured under the liaison (L-) designation from the 1950s on to the present time, all derivatives of civil aircraft designs, but the time of the very light liaison airplane of World War II was finished after the L-19 with missions being taken over by Army light helicopters. Planes originally in the category of liaison were heavier, more powerful, and often twin-engined, and in the 1962 aircraft designation re-ordering these aircraft were changed from liaison more properly to U- for utility plane. These included the DeHavilland L-20 becoming a U-6, the twin-engined Beech L-23 to U-8, the L-26 Aero Commander to U-4 and U-9, the twin engined Cessna 310 as Army L-27 to U-3, and the Helio L-28 to U-4. There have been no recent airplanes placed in the L- for Liaison category.

173.STINSON L-1 The initial airplane type classified in the liaison mission was the L-1 Vigilante, redesignated in 1942 from its original 1940 O-49 observation designation. Designed as a close-in Army cooperation aircraft with short-field performance, the L-1 wing was equipped with both automatic slots and slotted flaps. The 142 L-1s produced in 1940 and 1941 were converted to L-1B ambulance versions as predecessors of many other L-1C ambulance models. These two place 3400 pound aircraft were powered by a Lycoming R-680 radial engine of 295 horsepower and had a high speed of about 125 mph.

174.TAYLORCRAFT L-2A The Army L-2 Grasshopper, seen here in post-war rehabilitated form, was one of three pre-war civil lightplane designs procured in great quantity in militarized versions by the Army for battlefield liaison duties, the other two being Aeronca and Piper types. The L-2A was earlier designated an O-57A observation aircraft. A total of 1040 L-2, L-2A, and L-2B variants were procured; a further eight were drafted civil airplanes, and 900 more L-2M Army versions were delivered in 1943. Most all of these light planes were powered with 65 horsepower Continental horizontally-opposed engines, weighed between 1200 and 1300 pounds, and could attain a maximum speed of about 90 mph.

175.AERONCA L-3B The L-3 was the Aeronca version of the Army Grasshopper type, formerly designated a 1942 O-58 observation type, and then to the liaison category. As with the L-2, a substantial number (1439) were ordered directly as military L-3s, and a few (26) were impressed civilian planes. The L-3B photo shows one of 875 airplanes bought in 1942 powered by a 65 horsepower Continental engine, weighing 1850 pounds loaded, and having a high speed capability of 80 mph. Military versions had a wider fuselage and bigger windows than the civil Aeronca 65, upon which the design was based.

176.PIPER L-4H The photo shows an example of the Piper version of the Army Grasshopper light liaison airplane able to get in and out of restricted forward areas. The Piper was based on the famous pre-war civilian Piper Cub lightplane, and more examples of this Grasshopper type, formerly an O-59, were procured than any other. Fully 5554 Piper L-4s were delivered, not including 117 civil airplanes drafted in, with all of these entering service between 1942 and 1945. Usually powered by a 65 horsepower Lycoming engine, the militarized Cub could attain speeds up to 85 mph.

177.STINSON L-5E The Stinson L-5 military Sentinel was a development of the pre-war Stinson 105 Voyager, and when originally procured was labeled an O-62 observation plane. In 1942, 275 O-62s were ordered by the Army and, shortly, the program was expanded to cover an additional 1456 L-5 aircraft, this being the new designation. These were followed by an additional 1552 L-5B, C, E, and G Sentinels with minor changes; the one in the photo is a post-war civilian L-5E. The L-5s weighed 2200 pounds loaded, had a wingspan of 34 feet, and were powered by a Lycoming O-435 engine of 185 horsepower that provided a high speed of 115 mph.

178.BOEING YL-15 The YL-15 Scout liaison airplane was an unusual design from Boeing in Wichita for the Army. Two XL-15s and 10 YL-15s were delivered in 1946 and 1947. The observer had a particularly clear view, and extremely slow flight was obtained by use of a thick wing airfoil and unique full-span "flaperons" separate from the wing. Gross weight was 2110 pounds, wingspan was 40 feet, and the power of a 125 horsepower Lycoming engine provided a top speed of 104 mph.

179.AERONCA L-16A A 1947 update of the earlier L-3 Grasshopper, the L-16 was an Army liaison version of the civil post-war Aeronca Champion. The L-16A and B versions together totaled 609 aircraft, with many ordered for the National Guard. These later went to the Civil Air Patrol. Powered by a 85-90 horsepower Continental engine, the 1400 pound L-16 cruised at about 100 mph and had a high speed of 110 mph.

180.NORTH AMERICAN L-17A A 1947 military version of the commercial Navion, the L-17 was originally a North American design for a post-war sport plane. Ryan bought the design rights after North American produced 83 L-17As for the Army and Air National Guard, and added 158 L-17Bs. The Navion had a Lycoming horizontally opposed piston engine, providing a top speed of 150 mph at a gross weight of 3000 pounds.

181.CESSNA L-19A One of the most popular post-World War II liaison aircraft, the L-19 was a military modification of the civil Cessna 170. Over 3100 of the two place high wing monoplanes, labeled the Bird Dog, were produced through the 1950s and early 1960s. It was later redesignated to a new observation category in 1962 as an O-1 in a new series. Some production went to the U.S. Marine Corps as O-1E airplanes, and these were used in Korea. There were several versions of the Bird Dog: L-19A/O-1A to L-19E/O-1E. Typical characteristics included a 200 horsepower Continental engine, a gross weight of 2400 pounds, and a top speed of 150 mph. Bird dogs could operate as twin float seaplanes, and they were also exported to at least a dozen other countries.

182.DeHavilland L-20A The Canadian DeHavilland DHC-2 Beaver, a rugged bush-type single engine high wing monoplane convertible to twin floats, had a first flight in 1947, and in 1950 both the U.S. Army and USAF became interested in the aircraft and ordered six service test YL-20s. In the 1950s 968 L-20A aircraft had served both services, and many saw use in the Korean War. With the big aircraft redesignation in 1962, U.S. Beavers were changed to a U-6A designation; Beavers were also exported to at least 20 other countries. A Pratt and Whitney Wasp Junior engine of 450 horsepower gave the 5100 pound Beaver landplane a top speed of about 150 mph.

183.BEECH L-23D The L-23 six place twin engine light transport airplane for the Army was a militarized Beech Twin Bonanza that first appeared in 1952, named the Seminole. Production for the military through the 1950s totaled over 200 aircraft, principally in the L-23A, B, D, and F versions. They were widely used as staff transports throughout the world. The few L-23F final variants were based on the civil Beech Queen Air design. In 1962, all L-23 airplanes were redesignated as U-8 utility aircraft. L-23D airplanes had two Lycoming engines of 340 horsepower each, weighed 7000 pounds loaded, and had a maximum speed of about 235 mph.

184.AERO DESIGN L-26C Over two dozen Army and Air Force versions of the civil Aero Commander were purchased. The aircraft pictured is one of four L-26C Army airplanes, the equivalent of the civil 680 version. First flight of the early Commander was in 1958, and the design progressed through the 1960s. Powered by two Lycoming 290 horsepower engines, the handsome high wing Commander had a gross weight, depending on the version, of 6500 to 7000 pounds, a high speed of 250 mph, and could cruise at just over 200 mph. At least a dozen foreign countries also purchased the Aero Commander. In 1962 the Liaison designation was changed to the U-9 utility type.

185.CESSNA L-27A The Cessna L-27A was a military version of the popular commercial Model 310 twin piston engine four place aircraft. In 1957 the Air Force contracted for 160 L-27As, and later for 35 slightly refined L-26Bs corresponding to civil 310 development. A sleek low wing light transport with fixed wingtip-mounted fuel tanks, the L-27s weighed 4700 to 4800 pounds loaded, had a span of 36 feet, and used two Continental O-470 horizontally opposed piston engines each of 240 horsepower, allowing a maximum speed of about 230 mph. The 1962 redesignation system changed all L-27s to U-3s.

186.HELIO L-28A The photo shows a seaplane version of the Helio L-28A, later a U-10A starting in 1962. This Helio aircraft was an outgrowth of a STOL plane design of Otto Koppen at MIT in the late 1940s. A considerable quantity of U-10A and B airplanes were procured for special military operations and were used during the Vietnam War. Equipped with large wing flaps and full span wing slots, the STOL U-10B aircraft employed a Lycoming GO-480 six cylinder piston engine of 295 horsepower, spanned 39 feet, and had a gross weight of 3400 pounds. A six place airplane, the U-10B Super Courier had a high speed of about 165 mph. The Super Courier flew first in 1958.

OBSERVATION

The two place observation airplane was a stalwart of the US Army from World War I when that service, as the American Expeditionary Force in France, purchased 705 French Salmson Corps Observation aircraft in 1918 for use in that European war until 1941 when America entered World War II. As it turned out World War I was the last time US forces deliberately employed observation planes in battle until the Vietnam War.

In 1919 and for some time after the standard US Army observation plane was the American built DH-4, a British DeHavilland design powered by the American Liberty engine. Almost 5000 DH-4s were built during and just after World War I, and an almost infinite variety of modifications and updates of this aircraft type took place through the 1920s. A series of Corps Observation designs were put forth as prototypes by the Army Engineering Division and private companies from 1920 to 1924, but no production resulted from these efforts.

Perhaps partly as a result of success of planes like the British two seat Bristol Fighter in World War I Army post-war strategists apparently felt similar two seat observation aircraft could survive in a modern war. Observation planes were widely used and considered very important to effective ground operations. There were twice as many observation squadrons as pursuit or bomber squadrons, and over three times the number of attack squadrons.

In 1924 it was decided after aircraft prototype evaluations that the new Douglas two seat observation biplane was superior to the others, and this aircraft was ordered into production as the Liberty engined O-2, thus starting deliveries of over 200 aircraft of that type through the next five years. The competitive aircraft over this period, though with less Army orders, was the Curtiss O-1 Falcon usually with the Curtiss D-12 engine. The O-1 was slightly lighter and a little faster, but of generally similar characteristics. In 1927 Curtiss obtained additional orders with a Liberty engined O-11 Falcon, but through the 1920s the only other observation plane order was in 1928 for 29 Consolidated PT-3 primary trainers redesignated for some reason to O-17 observation planes for the Air National Guard.

In the 1930 to 1932 time period the Thomas Morse Company obtained orders for a total of 177 O-19 Wasp powered conventional biplane observation types, but with the advantage of having all metal fuselages. At the same time Douglas produced an update of their O-2 model as the O-25 design with a Curtiss Conqueror engine. Performance was somewhat improved over the O-2 with a 25 mile per hour increase in high speed to about 160 mph. Also in the early 1930s Douglas obtained contracts for another biplane, this a Hornet engine powered O-38 series with 156 delivered in several versions between 1931 and 1934. Curtiss obtained a consolation contract for a small number of a final observation Falcon in 1932 and built ten O-39s for the Army, this an updated O-1 with a Curtiss Conqueror engine.

In 1931 Douglas had experimented with high wing monoplane observation types as Army O-31 models with Conqueror engines, and three years later this test work paid off with a small production order for two dozen O-43 aircraft with a wire-braced high parasol wing giving excellent downward visibilty and a Conqueror engine. Douglas was good at building on a basic design and in 1936 delivered 90 refined O-43s with a strut braced high wing and a Twin Wasp Junior radial engine called the Army O-46. This cleaned up observation plane still with fixed landing gear could reach a 200 mph high speed, but this was about 100 miles per hour slower than modern fighter planes of the period.

Two more Army observation types obtained production orders in the years just prior to World War II, one a North American O-47 three place mid-wing monoplane of 1937 to 1939, and the other a high wing two place Curtiss O-52 Owl of 1940-41, each of which gained contracts for over 200 aircraft. The thinking of Army planners in the observation plane area seems fuzzy in retrospect. That an ungainly two or three place airplane with a high speed of little more than 200 mph and typically armed with a single rifle caliber fixed forward firing machine gun and another single flexible gun aft could survive in a then-modern world of 300 to 350 mph multi-gun fighter aircraft is a stretch of imagination. Luckily just before World War II started for the US the Army realized its errors and never intentionally let an airplane with the O- designation get into combat. Very light Army cooperation aircraft had their designations quickly changed from O- to L- for Liason.

The Army also found it useful to procure amphibian aircraft for observation work and in 1924 established a Corps Observation Amphibian category (COA-), soon changed to just Observation Amphibian (OA-). These amphibians were useful in overseas areas near water such as the Panama Canal Zone and the Philippine Islands where exploration and patrol flights were undertaken. In the mid- to late 1920s a total of 54 Loening-Keystone OA-1 and OA-2 two place amphibian biplanes (Loening sold out to Keystone prior to construction of the OA-2s) were procured by the Army. Landing gear wheels folded into the single main float which was integrated into the fuselage.. The OA-1s used an inverted Liberty engine and the OA-2s had the inverted Wright Tornado. These were the only OA- airplanes to carry armament, a single fixed gun forward and a rear flexible weapon.

The next Army observation amphibian types were military versions of the popular Douglas twin engine Dolphin redesignated OA-3 and OA-4 from their original cargo designations.

From 1933 to 1936 29 of these four place unarmed planes were used in the observation role by the Army, and a year later another amphibian was ordered by that service; five Sikorsky S-43s were procured as OA-8s. These graceful twin engined Hornet powered 11 place aircraft were employed for both personnel and cargo carrying. Then Grumman was the beneficiary of more orders for amphibians under the Army observation banner, putting out 31 twin engined G-21 Goose models as OA-9s in 1938 and 16 military versions of their smaller G-44 Widgeon in 1942 as OA-14s. But by far the largest orders for Army amphibians were for the OA-10 Catalina types throughout the World War II years with 361 of the big twin engined aircraft built, 131 by Convair and 230 by Vickers in Canada. These planes were useful for coastal patrol, rescue, and anti-submarine work, the OA-10 duplicating the Navy PBY-5A, the OA-10B the PBY-6A, and the OA-10A the Navy's PBV-1 also built by Vickers, Canada.

The last amphibian utilized by the USAF for many years, starting in the late 1940s, primarily for air-sea rescue missions, was the Grumman SA-16 Albatross. Under the older designation system the aircraft would have been designated as an OA- type, but since that category had been retired the Albatross was designated SA-16 for Search Amphibian, 16th type. (the OA-s had run up to -15). It was later in 1962 redesignated to a utility airplane , an HU-16. (H meant search and rescue!)

After a lapse of a few years, in the 1960s the Army got back into the observation plane business by procuring a large number of two seat twin engine Grumman OV-1 Mohawks, short takeoff planes able to use alternately photography, radar, and infra-red sensing and data linking information back to base. Though not armed with machine guns the Mohawk could carry various stores underwing, including rocket pods. The Mohawk was perhaps the only modern near-combat type airplane used by the Army.

Another late model lightly armed observation/reconnaissance airplane used by the Air Force in forward air control duties during the Vietnam War was the North American OV-10A Bronco first delivered in 1968 and not intended for a high intensity war.

It is difficult to follow all the twists and turns of Army and Air Force observation airplanes through the years with series of designators such as: CO-, O-, COA-, OA-, OV-, SA-, and HU-.

187.CURTISS R-2 In 1916 the Army Signal Corps procured a dozen Curtiss R-2 two seat reconnaissance airplanes for use in the Mexican War where they had many misadventures. A development of the basic Curtiss R design, this double-bay biplane aircraft had a Curtiss liquid cooled VX engine of 160 horsepower providing a maximum speed of 86 mph. The overhung upper wing of the R-2 spanned 48 feet and loaded weight was 2800 pounds. The curved bumpers under the lower wing tips protected them in a roll on the ground. The national insignia star on the rudder was red.

188.CURTISS R-4 The production version of the Curtiss R and, with the exception of the Curtiss JN, the most produced aircraft for the Army Signal Corps prior to US entry into World War I with 53 delivered by 1917. A heavier more powerful version of the Curtiss R-2, with a gross weight of 3240 pounds and a Curtiss V2-3 engine of 200 horsepower, the R-4 high speed was 90 mph. The Curtiss photo of an R-4 is dated December, 1916. R-4 airplanes were used as observation craft in the Mexican War.

189.BOEING DH-4M A 1924 manufacturer's photo shows a Boeing rebuilt DH-4M, the M standing for "modernized" with a steel tube fuselage and other improvements. Boeing in 1920 had rebuilt 111 wooden DeHavilland DH-4 Liberty Planes and later put out 187 DH-4Ms. Over 4700 DH-4s had been built by Dayton-Wright and Fisher Body during World War I, and the ubiquitous observation plane type with modernization served well into the 1920s. The Boeing DH-4M had a span of 42 feet five inches, a gross weight of 4600 pounds, and its 400 horsepower Liberty engine provided a high speed of 118 mph. Armament consisted of two fixed and two flexible machine guns, and 400 pounds of bombs could be carried.

C190.URTISS O-1B The Curtiss Falcon, along with Douglas observation types, was the middle-to-late 1920s replacement for the earlier DeHavilland DH-4. The first Army order for the Falcon was for 10 O-1s in 1925 using a Curtiss D-12 engine. O-1s were also tested with both Liberty and Packard engines. The O-1B shown in the photo was one of 25 purchased in 1927 and shows the external drop tank under the fuselage. The two place 4275 pound O-1B was powered with a Curtiss D-12 water cooled engine of 435 horsepower providing a high speed of about 140 mph. Armament consisted of one fixed and two flexible .30 caliber machine guns.

191.CURTISS O-1C This Curtiss photo of a low flying O-1C Falcon shows off the aircraft configuration well. In 1927 four O-1B observation planes were converted to VIP transports by making the rear cockpit wider than normal and thus more comfortable. In addition a headrest was mounted and armament was eliminated. In this photo the Army general is missing from the rear seat!

192.CURTISS O-1E The O-1E was the 1929-30 edition of the Curtiss Falcon line with modified ailerons and elevators, oleo shock struts on the landing gear, and armament changed to one fixed and one flexible gun. This January, 1930 company photo shows a new O-1E Falcon ready for delivery to the Army. Using the same engine as the O-1B, and with closely the same gross weight, the O-1E had similar performance. The order for O-1Es covered 37 aircraft , but two were converted to other Falcon observation variants for test. The O-1Es served in observation squadrons through the early and mid-thirties.

193.CURTISS O-1G In 1931 Curtiss delivered another batch of new Falcons, this model the O-1G, 30 of which were produced after testing a modified O-1E as a Y1O-1G. Little was changed except rear cockpit and flexible gun mount, revised nose lines, and modified landing gear. Gross weight rose to 4488 pounds, reducing climb capability with the same Curtiss D-12 engine. Speed was about the same and, like the O-1E, a single fixed forward firing gun and a single flexible gun comprised the O-1G armament.

194.DOUGLAS O-2H A direct competitor to the Curtiss O-1 Falcon, the Douglas O-2 was produced over approximately the same period, 1924 to 1930, but more than twice as many O-2 variants, O-2 to O-2K, were sold to the Army (242 Douglas to 105 Curtiss). The O-2H shown was the most produced, 73 aircraft being delivered in 1928. Unlike the Curtiss, the Douglas O-2 always used the Liberty engine of 435 horsepower. Gross weight ranged from 4700 to 4800 pounds and high speed was about 10 mph slower than the Falcon, around 130 mph. Armament was the same, a single fixed forward gun of .30 caliber and either twin flexible Lewis (on early models) or one flexible Browning (later models) in the rear.

195.DOUGLAS O-2H The photo shows an O-2H observation plane at Fort Sill, Oklahoma in 1929. The radiator for the Liberty 12-cylinder watercooled engine is at the nose right behind the propeller; radiator air exited through the cowl louvres. The flexible gun is absent from the rear cockpit ring mount. The O-2 had a straight upper wing as opposed to that of the O-1 Falcon. Wingspan was 40 feet and length 30 feet, a couple of feet greater in each direction than the O-1.

196.DOUGLAS O-5 A considerably larger and heavier observation airplane than the O-2, the five Douglas O-5s delivered to the Army in 1925 were equipped as twin float seaplanes. They were powered with the same 435 horsepower Liberty engine. The photo shows a near-equivalent Douglas DWC (Douglas World Cruiser) as a landplane, this airplane being the "New Orleans" in the collection of the USAF Museum and one of the actual planes used in the around the world flight project of 1924. The O-5s differed principally in having much less fuel capacity as well as incorporating normal military armament and other equipment. Span was 50 feet, seaplane weight and speed 9000 pounds and 107 mph respectively.

197.CURTISS O-11 In 1927 an O-1 Falcon observation plane was tested as an XO-11 with a Liberty 12A engine of 420 horsepower in place of the 435 horsepower Curtiss D-12. Two O-1A aircraft had been procured the previous year with the Liberty. So many Liberty engines had been produced that they were cheap. As a result Curtiss delivered in 1927 a total of 66 O-11 Liberty powered observation aircraft. The 4560 pound gross weight O-11s were the same size and had about the same horsepower as the D-12 powered O-1s. The company photo shows a new O-11 alongside a Curtiss Challenger engined Robin civil sport plane.

198.CURTISS O-11 This Curtiss photo shows a standard O-11 observation plane with a massive nose contour on the left and a Falcon with a smaller D-12 profile on the right. The main difference in powerplant depth was due to a change from a water cooling radiator on the O-11 to the then-new Prestone cooling radiator on the Falcon at the right. The O-11 Falcon with a water cooled Liberty engine did present a very large nose volume, and Prestone cooling was a major improvement.

199.CURTISS Falcon Demonstrator Another Curtiss photograph, though not of an Army aircraft, shows the large improvement in frontal area attained by use of a Prestone cooling radiator on a Falcon, this a demonstrator for export sales, principally in South America where dozens of Falcons were sold. The aircraft is a civil model belonging to Curtiss. Details of powerplant installation and split-axle landing gear with shock absorbers are seen clearly. A starter hand crank was stowed in an easily accessable position alongside the fuselage.

200.THOMAS MORSE O-19E The final product of the Thomas Morse Company which was absorbed into Consolidated was the O-19 two seat observation biplane. The O-19 became a standard Army type in the early 1930s with a total of 177 aircraft built and delivered between 1928 and 1932. The major distinguishing feature of the O-19 design was its all aluminum framing and corrugated aluminum fuselage covering. Only wing and tail surface coverings were of fabric. The final O-19E version shown in the photo was powered , like its predecessor versions, with a Pratt and Whitney Wasp engine uprated to 575 horsepower in the E. The O-19E spanned 40 feet, had a gross weight just under two tons, and with a ring cowl over the engine achieved a high speed of 156 mph.

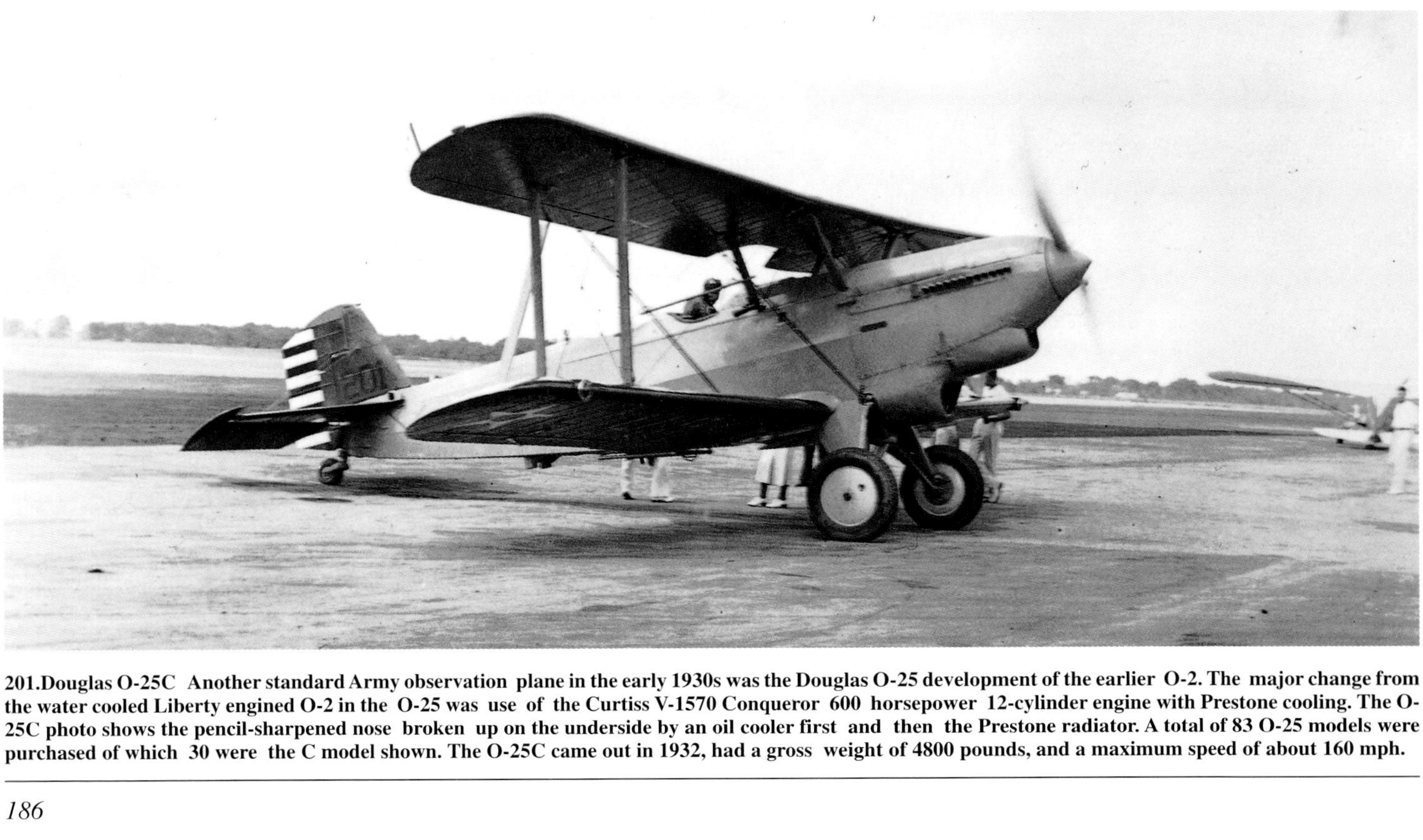

201.Douglas O-25C Another standard Army observation plane in the early 1930s was the Douglas O-25 development of the earlier O-2. The major change from the water cooled Liberty engined O-2 in the O-25 was use of the Curtiss V-1570 Conqueror 600 horsepower 12-cylinder engine with Prestone cooling. The O-25C photo shows the pencil-sharpened nose broken up on the underside by an oil cooler first and then the Prestone radiator. A total of 83 O-25 models were purchased of which 30 were the C model shown. The O-25C came out in 1932, had a gross weight of 4800 pounds, and a maximum speed of about 160 mph.

202.DOUGLAS YO-31 The two place O-31 design from Douglas was one of the first high wing monoplane observation types for the Army. From 1930 to 1932 13 O-31 aircraft were delivered in a variety of designations and configurations. The photo shows a YO-31, modified in 1931 from one of the two XO-31 prototypes of 1930. It was characterized by a corrugated aluminum sheet covering on its metal fuselage and a parasol wire-braced wing gulled into the upper fuselage. Armament consisted of a single fixed wing-mounted machine gun and a rear flexible gun, both .30 caliber. The 4650 pound gross weight YO-31 was powered by a Prestone cooled 600 horsepower Curtiss Conqueror engine giving a top speed of 182 mph.

203.DOUGLAS O-31B Another variant of the O-31 series, this O-31B was a single aircraft. The photo shows the central strut pylon support for wing wire bracing and the smooth metal skin covering the fuselage along with wheel pants and enclosed cockpits. The 1932 aircraft was quite similar to the five YO-31A models delivered the same year. Gross weight and engine power were the same as the YO-31. A graceful design, the O-31B signaled the Army's entrance into metal monoplane observation designs.

204.DOUGLAS O-31B A second view of the single O-31B minus wheel pants shows the clean nose contours of the inline engine installation broken up on the top by carburetor intake and below by oil cooler and Prestone radiator along with the wire bracing of the gull wing. The O-31B was provided to the National Guard as an unarmed VIP transport in Washington D.C. The O-31 airplanes were the direct predecessors of the later Douglas O-43 and O-46 high wing observation types.

205.DOUGLAS XO-35 Douglas and Fokker both produced a few experimental and service test three place twin engine high wing monoplane observation planes for the Army in the early 1930s. The photo shows the single XO-35 prototype of 1931 with the Wright Field arrow insignia on the corrugated aluminum covered fuselage. Five YIO-35s with a smooth metal fuselage followed for service test in 1933. The airplane was also tested as an XB-7 bomber. The Army decided they did not need the aircraft and there was no further production. The 10250 pound gross weight airplane used two Curtiss Conquerer engines each of 600 horsepower giving a maximum speed of 178 mph. The XO-35 had a wingspan of 65 feet eight inches.

206.DOUGLAS O-38B Douglas kept getting contracts for Army observation aircraft with its biplane O-38 series of 1932 following the earlier O-25 biplanes even while experimenting with monoplanes. This series was powered by the Pratt and Whitney R-1690 Hornet radial engine type. The O-38B airplane shown in the photo was one of 63 produced following an earlier run of 46 O-38s. There were few differences between these 109 aircraft. Many O-38 types were also exported. The O-38B used a 525 horsepower Hornet engine, weighed 4460 pounds loaded, and had a high speed of just under 150 mph. Armament consisted of one fixed and one flexible .30 caliber machine gun and up to 400 pounds of bombs underwing.

207.DOUGLAS O-38E Ordered for the Army National Guard squadrons, the next production version of the Douglas biplane observation type was the two place O-38E, a cleaned up model with a new faired fuselage, an uprated engine and three bladed propeller, along with enclosed cockpits. A total of 37 went to the Guard in mid-1933. Armament was similar to the O-38B model. The 625 horsepower Pratt and Whitney R-1690 Hornet engine provided a high speed of 154 mph for the O-38E. Wingspan was about 40 feet and gross weight 4600 pounds. The O-38E served through the 1930s and several were still in service in 1941.

208.Douglas O-38F Eight O-38F airplanes of 1934 capped off the Douglas observation biplane series; the new trend was all towards monoplanes, preferably with a high wing to prevent obstruction of downward view. The F models had dual controls as opposed to the E, but were otherwise very similar in characteristics and performance. They were most often used as VIP transports with no armament.

209.CURTISS O-39 The last of the biplane Falcon observation line for the Army, the Curtiss O-39 of 1931-2 was a cleaned up version using a small Prestone radiator under the Conqueror engine. Ten O-39s were ordered and except for export aircraft were the final Falcons. Wings and tail surfaces were essentially the same as earlier versions; Curtiss was noted for getting the final stretch out of a design, though Douglas obtained many more orders for biplane observation aircraft.

210.CURTISS O-39 One of the 10 O-39s fixed up as a special VIP airplane with wheel pants and enclosed cockpits. Many observation aircraft of the period were used for other than their primary intended mission. The Curtiss Conqueror engine was rated at 600 horsepower and was a liquid cooled vee type. Gross weight of the O-39 was 4650 to 4725 pounds depending upon equipment and maximum and cruise speed were approximately 175 and 150 mph respectively.

211.CURTISS YO-40A Following the trends of Army observation planes towards more all metal construction and less lower wing to block downward view, Curtiss put out the Raven in 1931-2 as a sesquiplane using a very small narrow lower wing, sort of half way between high wing monoplane and biplane. After the crash of the YO-40 in May, 1932 the aircraft was revised to a YO-40A as shown in the photo. This type featured mostly metal construction, an enclosed cockpit, and retractable main landing gear, the latter an innovation in observation plane design. The single YO-40A shown in the March, 1933 company photo had a Wright R-1820 Cyclone engine of 690 horsepower swinging a three bladed propeller, an all-up weight of 4740 pounds, spanned 44 feet, and had a high speed of about 185 mph.

212.CURTISS Y1O-40B The Curtiss Raven became a high wing monoplane with the Y1O-40B version of 1933 which came equipped with leading edge slots and trailing edge flaps. The lower wing was removed inboard to the landing gear fairing. Overall vision was thus improved. Weight went up slightly as did speed capability. In any case the Army ordered only four B versions, making a total of only six Ravens procured. The 4900 pound Y1O-40B was powered by a 690 horsepower Wright Cyclone giving maximum and cruise speeds respectively of 188 and 160 mph.

213.DOUGLAS Y1O-43 The O-43 observation type was developed directly from the Douglas O-31 series. In fact the five Y1O-43 service test airplanes of 1933 were YO-31Cs with the wing changed from an O-31 gull type fairing into the fuselage to an O-43 type parasol arrangement set above the fuselage on struts. The trim fuselage lines with the vee type engine up front are shown in the photo, but the wires and strut pylon above the wing are disguised by the wing in the photo. Wingspan was 46 feet; gross weight 5000 pounds.

214.DOUGLAS O-43A **Douglas obtained a production contract for the Conqueror-powered parasol wing O-43 type; in 1934 two dozen O-43A aircraft were delivered to Army observation squadrons. Changes included removal of propeller spinner, a roomier cockpit, and a revised engine exhaust. The photo clearly shows the parasol wing strut pylon and bracing wires. A little heavier at over 5300 pounds gross weight, the O-43A maximum speed at 191 mph benefitted somewhat from an uprated 675 horsepower Conqueror engine.**

215.DOUGLAS O-46A **The final Douglas high wing observation monoplane type came out in early 1936 first as an O-43A modified with a new wing and a Pratt and Whitney Twin Wasp Junior radial engine and then as 90 O-46A production aircraft. A large number of these airplanes went to the Air National Guard. The O-46A wing eliminated the upper pylon and wire bracing and used two struts on each side between fuselage and lower wing surface. In addition the cockpit retractable canopy faired directly into the fuselage contours when the rear gun was not in use. The O-46A was considerably heavier than the O-43A at 6135 pounds gross weight. The 725 horsepower Twin Wasp Junior got high speed up to 200 mph. Some O-46A airplanes lasted into 1941-2; a few were in the Philippines when the Japanese attacked.**

216.NORTH AMERICAN O-47A The all metal cantilever wing monoplane with retractable landing gear finally came to Army and National Guard observation squadrons in the form of the North American O-47, a three place type with the centrally located observer able to see out windows below the mid-wing. The standard armament of one fixed and one flexible .30 caliber machine gun was employed. After one XO-47, a General Aviation design, North American rolled out no less than 164 O-47A and 74 O-47B models in the 1937-9 time period. Powered by a Wright Cyclone engine of 900 to 1000 horsepower, the O-47s weighed from 7650 to 8100 pounds loaded and could achieve a high speed of about 225 mph.

217.CURTISS O-52 The final "heavy" observation plane for the Army prior to World War II was the Curtiss O-52 Owl. Curtiss won a contract for 203 of these all metal high wing two place monoplanes in late 1939 and deliveries were made in 1941 to Army observation squadrons along with a few to Russia. Neither the US Army nor the Russians had much use for the type, and it was not used operationally, though some were lost early in the Philippines. With a gross weight of about 5300 to 5600 pounds the O-52 with its 600 horsepower Pratt and Whitney Wasp engine had a high speed of just over 200 mph.

218.CESSNA O-2A An unusual configuration set this O-2 military version of the civil Cessna Skymaster twin engine aircraft apart. First flight of a two to six place dual control O-2A took place in 1967, and the US military ordered 510 of the A and B models as forward air controller and psycological warfare types respectively. The 4630 pound aircraft was powered by one tractor engine and propeller forward and one pusher engine and propeller combination aft on the short fuselage, thus providing symmetrical engine-out safety. The two Continental six cylinder horizontally opposed piston engines were each rated at 210 horsepower and provided a maximum speed of 206 mph, Four underwing stores pylons could carry rockets, flares, or other light ordnance.

219.LOENING OA-2 For many years the US Army procured and operated observation amphibians (the OA- series) for coastal patrol, part of the Army's mission. One of the earliest of the type was the Loening OA-2 of 1929. After procuring several generally similar Loening OA-1s a few years earlier, the Army ordered an XOA-2 (later redesignated XO-10) and then eight OA-2s. The single main float two place biplane , an example of which is shown in the photo, had wheels retracting into the float. The OA-2 weighed 5400 pounds loaded, had a wingspan of 45 feet, and an inverted vee type Wright V-1460 engine of 480 horsepower gave the OA-2 a top speed of 112 mph.

220.DOUGLAS OA-4 The popular early 1930s Douglas Dolphin civil amphibian was purchased by the Army in 1932-3 as the Y1C-21 and Y1C-26 transport type, a total of two dozen airplanes being procured. Very shortly they were redesignated as OA-3 and OA-4 observation amphibians in five sub-variants. Pictured is one of four OA-4Bs. The four place twin engine aircraft mounted two Pratt and Whitney Wasp Junior engines each of 350 horsepower on struts high over the wing. The 8600 pound gross weight Army Dolphin had a high speed of 140 mph.

221.GRUMMAN OA-9 The Army kept buying observation amphibian aircraft, sampling just about every civil amphibian type produced including this six place Grumman G-21A Goose of which 31 were procured in 1938. The Army Goose weighed just under four tons fully loaded, had a wingspan of 49 feet, and was powered by two Pratt and Whitney Wasp Junior radial engines each of 450 horsepower cleanly mounted on the high wing. The standard Grumman type of retractable landing gear was used. The Goose had a top speed of 190 mph.

222.GRUMMAN OA-14 In 1942 the Army continued buying observation amphibians with the purchase of 16 five place Grumman OA-14 aircraft, a military version of that company's civil G-44 Widgeon. The photo shows one of these post-war, well worn and lacking propellers. Many were rehabilitated and re-engined after the war. The OA-14 had a 40 foot wingspan, weighed 4500 pounds loaded, and used two Ranger L-440 engines of 200 horsepower each that provided a maximum speed of 150 mph.

223.GRUMMAN OV-1D The Grumman OV-1 Mohawk was a specialized two place twin engine short takeoff and landing (STOL) airplane used in tactical observation and reconnaissance roles. It featured side-by-side seating and turboprop engines. Production started in 1960 and the last delivery was in December, 1970 with a total of 380 Mohawks produced. Various versions, OV-1A, B, C, and D were used depending on installation of photographic, side-looking radar, or infra-red sensors. Mohawks were used in the Vietnam War. The OV-1D weighed 18448 pounds loaded, spanned 48 feet, used two Avco-Lycoming T53 engines of 1400 equivalent shaft horsepower each, and had a maximum speed of about 288 mph depending on external stores carried. Two wing store stations could support 150 gallon drop tanks.

224.NORTH AMERICAN/ROCKWELL INTERNATIONAL OV-10A The winner in 1964 of a design competition for a light armed reconnaissance aircraft (LARA) to be used in counter-insurgency missions, the OV-10 Bronco had two of most everything: crew (in tandem), engines, tail booms, and vertical tails. First production flight of an OV-10A came in August, 1967, and after production ended in early 1969 the USAF had 157 in service (the USMC had 114 more). Some were used in Vietnam for forward air control. Several other versions went to foreign governments. The OV-10A had a high speed of about 280 mph and an overload takeoff gross weight of about 14500 pounds. Engines were two 715 equivalent shaft horsepower Airesearch T76 turboprops and propellers were three blade Hamilton Standard models.

225.GRUMMAN SA-16B Utilized by the USAF as an air search and rescue amphibian, the Grumman G-64 Albatross first flew as a Navy XJR2F-1 utility transport, but the Air Force MATS ordered 241 of the twin engined boat hull and tip float amphibians with retractable landing gear. In the late 1950s the SA-16A models were updated to Model 111 with new wings and tail and designated SA-16B. With the major service redesignation system of 1962 the Albatross became the HU-16A and B. Powered with two Wright R-1820 Cyclone piston engines each of 1425 horsepower, the 37500 pound gross weight aircraft had a high speed of about 235 mph. Wingspan was 96 feet eight inches after the modification. Drop tanks for fuel could be mounted underwing. SA-16Bs were used during the Vietnam War.

PRIMARY TRAINER

The primary trainer is the first plane into which a new pilot candidate steps. In some cases early on it would not even be designed to leave the ground. In 1918 the Army bought 301 Breese Penguins, a single place monoplane ground trainer with a wingspan of only 14 feet. Other ground trainers were purchased from the French by the American Expeditionary Force in World War I.

In the U.S. Army during late World War I and for some time thereafter the two primary training airplanes were the Curtiss JN Jenny and the Standard SJ-1, both two place biplanes having tandem seating, with many hundreds of each purchased by the Army Signal Corps. Well over 4000 JN-4 and JN-4D Jennies and at least 1600 Standards were procured. These were the primary trainers of 1917 and on into the 1920s.

In the early 1920s the Army tested several primary trainer candidates, and in 1922-23 ordered 13 new Dayton Wright TA-3s and 20 Consolidated TW-3s. The A and the W stood for Aircooled and Watercooled engine equipped respectively. Interestingly, these light biplanes both had side-by-side seating arrangements for instructor and trainee, a feature that was later discarded for tandem seating and then picked up again many years later in the Cessna T-37A jet trainer.

In 1925 the first of the "modern" series of Army primary trainers came out as the Consolidated PT-1, just a revision of their earlier TW-2 to tandem seating. Consolidated, along with subsidiary Fleet (named for Reuben Fleet, head of Consolidated) dominated the Army market for primary trainer aircraft into the early 1930s, supplying a total of 527 aircraft of the PT-1, PT-3, PT-6, and PT-11 two place tandem seat biplane series from 1925 to 1932. Other PT-prototypes totaled only nine aircraft.

In 1936 Stearman broke into the Army primary trainer market with their PT-13, itself a development of their earlier PT-9 military trainer derivative of the civil Stearman Cloudboy. There were so many Stearman PT- biplane tandem seat trainer airframes produced that from 1936 through 1942 the output of three different radial engine manufacturers was required to power them. The 691 PT-13 models in five variants had Lycoming engines; the 3519 PT-17 models had Continentals, and the 150 PT-18s had Jacobs engines. These primary trainers, along with a very few military versions of the Waco YPF-7 civil aircraft as the PT-14, were the last of the biplane primary trainers except for the PT-27 noted later.

From then on through 1941-42 there were only two more primary trainers of note in the wartime period. The Fairchild PT-19/PT-23/PT-26 series low wing tandem seat monoplane mostly of wood construction was built in the thousands; the -19 and -26 with the Ranger engine and the -23 with a Continental radial. The Ryan militarized civil model ST all metal low wing monoplane was first a PT-16 or PT-20 with a Menasco engine and then a PT-21 or PT-22 using a Kinner radial engine, the latter proving much more satisfactory.

The Stearmans, Fairchilds, and Ryans were the big three primary trainers of World War II. Some time before the U.S. entered the war a plan to train 12000 pilots a year was instigated and such a training rate was attained in March, 1941, a few months before Pearl Harbor. This rate was attained using 28 civilian contract flying schools. Before that, primary training was done at Randolph Field in Texas. Shortly afterwards, under the pressure of wartime events, a new plan for training 30000 pilots a year was put into place, and 41 civilian contract schools were to be used.

The primary stage of training lasted ten weeks at that time. One primary trainer was maintained for every four to five cadets. Each cadet got about two hours of flying daily. Under the contracts the Army loaned the aircraft and the civilian school maintained them at night after flying ceased. This was the Civilian Pilot Training Program (CPTP). Army officers

352.CURTISS JN-4A An early version of the famous two place Jenny trainer, the name apparently coming from a combination of the early Curtiss models J and N. Records indicate that some number between one and 150 of this version were built in 1917 for training purposes. The aircraft pictured had a tilted down thrust line, ailerons on the lower wing, and a large amount of wing dihedral. Powered by a 90 horsepower Curtiss OX-5 engine, the 1870 pound gross weight JN-4A had a top speed of 75 mph. Wingspan was 43 feet seven inches and length 27 feet three inches.

353.CURTISS JN-4D The major production version of the OX-5 powered Jenny, the JN-4D version was produced in the hundreds by both Curtiss and at least seven other companies in 1917-18, and thus was easily the most utilized primary trainer in the U.S. during America's time in World War I and for some time afterwards. Powered by the OX-5 engine, the performance and size of the approximately 2000 pound JN-4D was similar to the JN-4A with a high speed of 75 mph.

supervised the work. In 1941 the primary training phase covered 60 flying hours and 141 hours of classes. About 60 percent of the trainees completed the course satisfactorily; three quarters of those eliminated were dropped because of a lack of coordination.

By 1943 the U.S. Army had all the primary trainers it needed for the war effort. The final aircraft designated in the PT- aircraft series was the Boeing PT-27 which was simply a PT-17 with a cockpit canopy for cold weather flying in Canada; 300 of this model were built. The number series was picked up post-war in the North American T-28A, but that aircraft was not a primary trainer.

The next primary trainer for the USAF was the Beech T-34A, an aircraft covered under the T- for trainer section elsewhere.

353-A.CURTISS JN-4H The JN-4H was different from the D version in using the Wright Martin Company version of the Hispano A engine delivering 150 horsepower, a very significant increase in installed power, and maximum speed went up to 93 mph with gross weight climbing to 2150 pounds. The H version nose radiator was rounded on the bottom and larger than that of the D model. Several versions of the JN-4H were produced to adapt for gunnery, bombardment, and observation training.

354.CURTISS JN-4H The photo shows a truly classic flight view of the JN-4H "Hisso Jenny" in Army colors of the day with the national insignia prominently displayed inboard on the wings and vertical striping on the rudder. Over 900 JN-4H Army two place biplanes were produced in 1918. These aircraft could fly about 250 miles and had a ceiling of about 7000 to 8000 feet.

355.CONSOLIDATED PT-1 In 1924 the Army revised its designating system for aircraft, and the first primary trainer using the then-new system was the Consolidated PT-1 Husky, a derivative of the earlier Dayton Wright TW-3 trainer. A total of 221 tandem seat single bay biplanes were procured in 1925 powered by a Wright E powerplant of 180 horsepower, an American version of the Hispano Suiza, an eight cylinder (in two banks of four) vee-type water cooled engine. Gross weight was 2575 pounds and high speed 92 mph. The PT-1 had a fabric-covered steel tube fuselage. Wingspan was 34 feet nine inches.

356.CONSOLIDATED PT-3A After an experimental prototype XPT-3 in 1927, which was a PT-1 with a radial engine, a new tail, and wings of Clark Y airfoil section, Consolidated received a contract in 1928 for 130 PT-3 trainers. Size and weight of the PT-3 were similar to that of the PT-1; the big difference was use of a Wright J-5 Whirlwind nine cylinder air cooled radial engine of 225 horsepower, providing a maximum speed of 125 mph. In 1929 the Army liked the PT-3 well enough to order 120 more aircraft as PT-3As with very minor changes.

357.STEARMAN YPT-9B In 1931 Stearman entered an Army competition for a new primary trainer with their commercial Cloudboy two seat biplane, and four were purchased by the Army as YPT-9s with a Wright J-5 Whirlwind engine of 165 horsepower. The next year the four planes were equipped with different engines, one a Continental powered YPT-9A, one Kinner powered YPT-9C, and two YPT-9Bs with Lycoming R-680 nine cylinder radial engines of 200 horsepower. The type was not ordered in quantity. Shown in the photo is a civil Model 6 Cloudboy painted up in Army colors to simulate one of the YPT-9Bs. The Stearman had a wingspan of 32 feet, weighed 2750 pounds loaded, and had a high speed of about 110 mph.

358.CONSOLIDATED PT-11D In 1931 and 1932 the Army, after ordering four Y1PT-11 and five PT-11B primary trainers, decided to give Consolidated a small production order for 21 PT-11D aircraft, keeping this company as the premier supplier of the type to the Army up to that time. Three of the Y1PT-11s and five PT-11Bs were also updated to D standard with use of the 200 horsepower Lycoming R-680 engine. The 2600 pound gross weight PT-11D had a maximum speed of about 117 mph. Some of these airplanes were redesignated as BT-7s and used as basic rather than primary trainers.

359.STEARMAN PT-13B Developed from the Stearman private venture X70 of 1934, the Stearman PT-13 series started off a long line of very successful primary trainers used by the Army. After producing 26 PT-13s and 92 PT-13As with Lycoming engines Stearman, in 1940, started production of the biggest lot of primary trainers yet, 255 PT-13Bs. The major change in the 2700 pound biplane was use of an uprated Lycoming R-680 radial engine of 280 horsepower, giving the 32 foot two inch span plane a high speed of 130 mph.

360.STEARMAN PT-13D With the advent of U.S. entry into World War II, it became desirable in 1942 to standardize primary trainers used by both Army and Navy aviation cadets. This objective was achieved in the PT-13D model of the Stearman trainer, 318 of which were procured by the Army to add to earlier PT-13 models; in addition, 1450 of the same aircraft went to the U.S. Navy as N2S-5s. Powered by a de-rated Lycoming engine of 220 horsepower, the PT-13D had a maximum speed of 125 mph. These PT-13 aircraft, along with PT-17s, were the first planes most new aviation cadets stepped into to start flight training.

361.WACO YPT-14 One of only 15 PT-14 aircraft purchased by the Army, this service test YPT-14 was a militarized two place version of the civil Waco UPF-7 biplane. Although just a few of this type were procured, many others in civilian UPF-7 form were utilized in the extensive Civilian Pilot Training Program (CPTP) during World War II. Powered by a Continental R-670 radial engine of 220 horsepower and weighing 2650 pounds loaded, the YPT-14 had a high speed of about 138 mph.

362.STEARMAN PT-17 Recipient of the largest Army order ever for a primary trainer, the Stearman PT-17 was like the PT-13 except for use of a Continental R-670 radial engine of 220 horsepower. It took the combined output of Continental, Lycoming, and Jacobs to provide engines for Stearman primary trainers being mass produced for the expansion of the U.S. air services. The PT-18 was the Stearman version using the Jacobs engine, an R-755 radial of 225 horsepower. Performance and weights were similar for all versions. The photo shows a civil post-war restored PT-17 at a 1975 airshow.

363.FAIRCHILD PT-19 The first monoplane to be put into production as an Army primary trainer, the Fairchild Model M-62 was given the basic designation of PT-19 in 1940 when a first batch of 270 aircraft was ordered. To meet projected wartime needs in the next two years, additional Fairchild PT-19A and B models with minor changes were produced by the Aeronca and St. Louis companies. The total of PT-19 types reached 4889 airplanes, more than any other PT design. A two place open tandem cockpit low wing monoplane with a gross weight of 2550 to 2650 pounds, the PT-19 was powered by a Ranger 200 horsepower air cooled six cylinder engine, giving it a high speed of about 130 mph.

364.RYAN PT-20 A military primary trainer adaptation of the popular Ryan ST-A low wing two place all metal sport monoplane, the PT-20 was a minor revision of the earlier 16 XPT-16 and YPT-16 Menasco powered test aircraft of 1939 to 1941. A total of 30 PT-20s were delivered in 1941 with the same 125 horsepower Menasco four cylinder Pirate engine, giving the 1635 pound trainer a top speed of 132 mph. When the Menasco engine failed to measure up to rough Army training use, a 132 horsepower Kinner radial engine was substituted in PT-16s and PT-20s with closely similar airplane performance and greater engine reliability, though the appearance of the airplane suffered.

365.RYAN PT-22 The major Army production version of the Ryan ST trainer, now known as the ST-3, the PT-22 dispensed with landing gear fairings and other niceties and used an uprated 160 horsepower version of the Kinner radial engine, giving a maximum speed of 125 mph. The appearance of the aircraft was not at all enhanced by fairing removals or use of the uncowled radial engine, but as a practical Army primary trainer it was an improvement over earlier models.

366.RYAN PT-22 A post-war civil rehabilitated PT-22 taxis out for a takeoff in January 1975 adorned in World War II markings. A total of 1023 PT-22 trainers were produced by Ryan in 1941. Twenty-five PT-22As were reclaimed by the Army from a Dutch contract for ST-3s powered by the 160 horsepower Kinner engine. The PT-22 had a 30 foot wingspan and weighed 1860 pounds loaded.

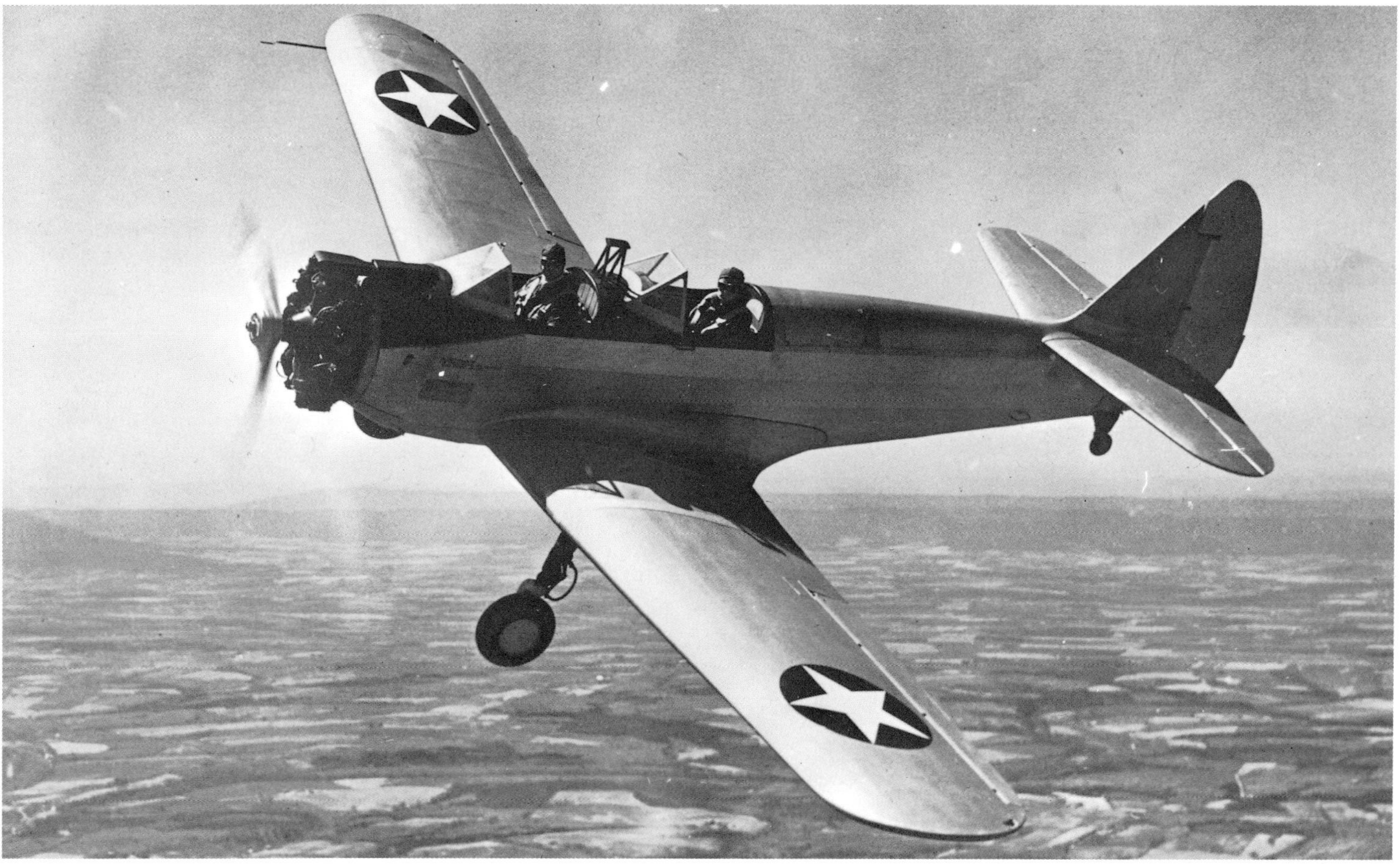

367.FAIRCHILD PT-23 With production of Fairchild airframes outrunning Ranger engine production, another engine was needed to power the Fairchild trainer. In 1942 a PT-19A was revised with a Continental seven cylinder radial engine of 220 horsepower as an XPT-23. Shortly, production was started by Fairchild and four other companies, Aeronca, Howard, St. Louis, and Fleet, on PT-23 and PT-23A aircraft, the latter equipped for blind flying training. In all, 1125 PT-23 type primary trainers were put out. Gross weight was about 2800 pounds and high speed about 125 mph.

368.NEG1085N FAIRCHILD PT-26A This version of the PT-19 was equipped with a canopy over both tandem cockpits to protect occupants from the severe Canadian winter cold. In 1942 Fairchild built 670 PT-26 aircraft so equipped. Fleet of Canada in 1942-3 built 807 PT-26A and 250 PT-26B trainers, also with a canopy. All PT-26s were powered by a Ranger L-440 engine of 200 horsepower, weighed 2750 pounds gross, and had a top speed of 128 mph.

RACING

In the early 1920s the Army procured racing planes or modified service or experimental types into racers to compete in national air racing events and thus put its prestige on the line, particularly against that of the Navy.

The U.S. Army Air Service entered a Verville VCP-R biplane, a modification of an experimental VCP-1 pursuit, with a 640 horsepower Packard engine in the Gordon Bennet Cup Race held in September 1920 near Etampes in France. The aircraft did not finish the race as it was forced out by an overheating engine, but in November of that year the Verville, later called an R-1 in a racing category, was flown in the first Pulitzer Race on Long Island, New York, by Capt. Corliss Mosely and took first place at an average speed of 156.5 mph.

In 1921 the Army entered a converted Thomas Morse MB-6 pursuit type as an Army R-2 racer in the 1921 Pulitzer Race at Omaha, Nebraska, piloted by Lt. J. Macready. The aircraft finished in third place with the race being won by Bert Acosta in a Navy Curtiss racer.

A year later in October 1922 at Detroit the Army entered several aircraft; three were Verville Sperry R-3 low wing monoplane racers flown by Army Lts. Barksdale and Johnson and Capt. St. Clair Street, along with two Curtiss R-6 racer biplanes flown by Lts. Maughan and Maitland. The two R-6 aircraft placed first and second with Lt. Maughan the winner, both beating out two Navy Curtiss racers. One of the R-3s came in fifth. Two Loening monoplane racers designated as R-4s were also in the 1922 Pulitzer, as were two Thomas Morse TM-22 all metal parasol monoplanes as Army R-5s piloted by Capt. Hunter and Lt. Bissell. The Loenings were later discarded.

In the 1923 Pulitzer Race at St. Louis the Army got soundly beaten by the Navy Curtiss and Wright racers. The best the Army could do was come in fifth in a Curtiss R-6 piloted by Lt. Miller.

The next year, 1924, saw only four entrants in the Pulitzer, all of them Army racers. A Curtiss R-8 that was to have flown in the race crashed beforehand. Two R-6s were flown by Capt. Bert Skeel and Lt. Brookley. The Verville Sperry R-3 entered after having dropped out of the 1923 Pulitzer, and was piloted by Lt. H. Mills. A Curtiss PW-8A modified Hawk was also entered, flown by Lt. Stoner. Diving starts were allowed in the race, and tragedy occurred when Capt. Skeel's R-6 lost its wings and crashed at the start. The Verville Sperry won the race at an average of 215.7 mph.

The Pulitzer Race of 1925, the last of the series, was held at Mitchell Field, Long Island, New York. As usual it consisted of four laps around a 31 mile race course. The planes took off at intervals short enough so they were all on the course at the same time, but they were all timed separately. The entrants were two new Curtiss R3C-1 racers (a Navy designation the Army did not bother to change), one flown by Army Lt. Cyrus Bettis and another flown by Navy Lt. Al Williams. Three standard Army pursuits were also entered. The Army R3C-1 won the race at 249 mph with the Navy R3C-1 second.

After the 1925 race Army participation in racing ceased and there were no more Pulitzers, and thus ended the short era of Army racing and the design of Army racing planes.

369.THOMAS MORSE R-2 One of the first Army racers, the R-2 was a 1921 conversion of one of three Thomas Morse MB-6 pursuit aircraft to a racing model. A very small airplane with a wingspan of only 19 feet, the R-2 used a special Wright Hispano H engine with power boosted from 300 to 400 horsepower. The R-2 was entered into the 1921 Pulitzer Trophy Race and placed third at an average speed of 160.7 mph behind a Navy Curtiss R-1 and a Curtiss Cactus Kitten triplane. The little Thomas Morse biplane had a gross weight of just over 2000 pounds.

370.CURTISS R-6 Two R-6 Army racers were built by Curtiss for the 1922 Pulitzer Race. A very trim little biplane with an upper wing skin cooling system for the engine and a 460 horsepower Curtiss D-12 powerplant, the R-6 racer type weighed 2120 pounds and had both a wingspan and length of 19 feet. The two R-6s ran away with the 1922 race with Lt. Maughan first averaging 205.8 mph and Lt. Maitland second averaging 198.8 mph, beating two Navy Curtiss racers and a Verville Sperry R-3. In 1923 an R-6 placed fifth and last at 218.9 mph in the Pulitzer of that year. It did not place in the 1924 race, having lost its wings in a diving start.

371.CURTISS R-8 In 1924 Curtiss built two aircraft for racing as Navy R2C-1s intended for use in the 1923 Pulitzer race. One of the two was purchased by the Army and designated an R-8, but the racer crashed before it could be entered in the 1924 Pulitzer. The R-8 was slightly larger than the earlier R-6 with a wingspan of 22 feet and a length of 19 feet nine inches, but was only 30 pounds heavier. The uprated Curtiss D-12 engine with skin cooling provided 500 horsepower and gave the R-8 a top speed of 266 mph, making it the fastest of the early Army racers.

372.CURTISS R3C-1 This racer was a joint Army-Navy effort along with Curtiss, one of three articles, and could be converted into a twin float seaplane. Though an Army airplane, it kept its Navy R3C-1 designation, and in 1925 the Army R3C-1 won the Pulitzer Race piloted by Lt. Cyrus Bettis (shown in the photo) at an average speed of 248.98 mph, just beating out a Navy R3C-1 piloted by Lt. Al Williams. The 2150 pound R3C-1 was an improvement of the R-8 using a Curtiss V-1400 engine of 619 horsepower. The upper wing skin cooling radiators can be seen clearly in the photo.

RESEARCH

The purpose of research aircraft generally has been to explore new regimes of flight or new design concepts for potential application to operational military aircraft. The designations for these types are somewhat confusing. The X- series (formerly XS- series) was the first category started just after World War II with the X-1 (XS-1 for only a short while). There were also two other series, V- and VZ-, for vertical and short takeoff machines.

The Bell X-1 series of airplanes became the first research aircraft and were used to explore flight at very high speeds and altitudes starting in 1946. They were highly successful overall. The X-1 made history by being the first aircraft to exceed the speed of sound in level flight in 1947. The Bell X-2 aircraft of 1952 were intended to further expand the speed-altitude envelope using a rocket engine—capable of using a throttle—for propulsion.

Other early research types were the single Douglas X-3 of 1952 intended for further high speed research using jet engine power for longer sustained flight; the Northrop X-4 Bantam, flight tested in 1948 for research of tailless swept wing aircraft configurations; and the Bell X-5 jet aircraft of 1951 built to explore the effects of various wing sweep angles, and a copy of a German design.

Lesser known types included the X-6, a never-realized B-36 version using nuclear power; the X-7 of 1951, with several built to test ram-jet engines; and the X-8 through X-12 test vehicles for various missile programs.

The next two X- models of the mid to late 1950s were test vehicles for vertical takeoff and landing (VTOL) aircraft concepts, the X-13 being something like a jet powered tailsitter but actually launched and retrieved near a vertical platform, and the Bell X-14, a vectored jet thrust research machine.

The North American X-15s were very successful in expanding aircraft flight research right to the edges of space with speeds approaching those of later spacecraft during the 1960s. The Lockheed X-17 test vehicles conducted research on nose cones designed for space vehicle re-entry into the atmosphere.

The Hiller X-18, Curtiss X-19, and Bell X-22 were 1960s VTOL aircraft respectively exploring the turboprop tilt wing, shaft turbine powered tilt propeller, and shaft turbine powered tilt ducted fan concepts. The first two accomplished a minimum amount of flight test work; the third had a long flight test program.

The Boeing Dyna-Soar orbital glider concept claimed the X-20 designation, and the X-21 was a Northrop modification of the Douglas B-66 bomber testing new wings employing an active laminar flow control system.

The Martin Marietta X-23 was an unmanned lifting body research vehicle, and the X-24, built in an A and B version, was a manned powered lifting body research plane of the early to mid-1970s.

In the V- series of V/STOL airplanes the McDonnell XV-1 of 1955 was a winged rotor-unloaded convertaplane with tip rotor drive, and the Bell XV-3 a mid-1950s pioneer fixed wing tilt rotor vehicle; both aircraft were powered by piston engines.

The XV-4 of the early 1960s was a V/STOL research aircraft first using the ejector jet principle, later a vertical jet lift type designed at Lockheed, and the XV-5 of the mid-1960s was a Ryan/General Electric jet powered research vehicle with two vertical lift ducted fans, one buried in each wing. Ryan had the airframe and General Electric the lift and propulsion system. The Army and NACA both tested the XV-5.

The XV-7 designation covered three Hawker P-1127 VTOL jet fighters of the 1960s imported for evaluation; these aircraft were prototypes for the later AV-8 Harrier used now by the Marines.

The two Bell XV-15 tilt rotor vehicles were used for a continuation, starting in 1973 with a contract award from the Army and NASA, of research into feasibility of that type of V/STOL machine. Work was sufficiently successful to pave the way for later construction of the Bell Boeing V-22 Osprey tilt rotor aircraft with its first flight in early 1989. The V-22 is intended for operational use by all branches of U.S. services; by January, 1996 V-22 test aircraft had flown more than 1076 hours in more than 940 flights demonstrating both vertical and horizontal flight.

The VZ- series of test vehicles was sponsored by the Army Transportation Research Command (TRECOM) starting in the 1950s. The VZ-1 was a Hiller design of 1959 employing a one place flying platform supported by a ducted propeller. The Vertol VZ-2 of 1957 was the initial test vehicle for the tilt wing turbopropeller V/STOL aircraft concept and was tested through the transition range between vertical and horizontal flight. Both Vertol and NASA conducted extensive testing which led later to the program for the Tri-Service Vought/Hiller/Ryan XC-142 tilt wing cargo transport design of the 1960s.

The Ryan VZ-3 and the Fairchild VZ-5 of the 1950s were small vehicles for testing the propeller deflected slipstream concept of V/STOL aircraft.

The Doak VZ-4 two place vehicle of the late 1950s was designed to explore a V/STOL concept embodying a fixed wing with a tilting ducted fan at each wingtip.

In the late 1950s the Army decided to investigate the idea of vertical lift "flying jeep" vehicles, and awarded contracts to Chrysler for a ducted propeller type, Aerophysics Division of Curtiss Wright for a four lifting propeller version, and to Piaseki for a design hav-

391.BELL X-1 The Bell X-1 series of thin straight wing research aircraft were produced with the purpose of exploring the supersonic flight regime. Three aircraft were built; the first X-1 was air-launched without power from a B-29 in January 1946, with powered flights starting the following December. In October of 1947 the X-1 shown in the photo made history by becoming the first aircraft to fly above the speed of sound, reaching Mach 1.015 (670 mph at 42000 feet) while piloted by Captain Chuck Yeager. Powered by a Reaction Motors 6000 pound thrust rocket engine, the X-1 had a gross weight of 13400 pounds and a maximum speed of 950 mph. The X-1 made a total of 156 flights.

ing two ducted rotors. After testing, no contracts for production were awarded.

Though not designated as research vehicles, five Vought Hiller Ryan XC-142A aircraft were built and tested in the early to mid-1960s under a Tri-Service program, and optimistically designated in the cargo transport series. They were to demonstrate all-weather operational suitability of a V/STOL tilt wing turboprop transport concept. Testing did not result in production.

392.BELL X-1B The X-1A had a new fuel pump system, a longer fuselage, and a revised cockpit canopy, as well as more fuel capacity. This aircraft broke more records in December of 1953 by making a speed of Mach 2.44 and, in June 1954, achieved an altitude over 90000 feet. The X-1A was destroyed after 21 flights. The X-1B, shown in the photo, was used for research into thermal heating effects at very high speeds and made 27 flights. The X-1C design was canceled, while an X-1D was destroyed after one flight in the summer of 1951 when it was jettisoned after an explosion in the B-50 mother plane.

393.Bell X-1E The photo shows an X-1E version, the second of the original X-1 aircraft modified with a thinner wing and a new windshield, and a new rocket control system for flight substantially outside the earth's atmosphere. The X-1E made a total of 26 research flights.

394.BELL X-2 The Bell X-2 was a swept wing research aircraft used to explore supersonic flight. Two aircraft were built by Bell using a considerable amount of stainless steel structure and powered by a throttle-capable Curtiss Wright liquid propellant rocket motor rated at 15000 pounds of thrust. The first plane was lost in 1954 in an emergency jettison from a B-50 mother aircraft. The second plane first flew in late 1955. On its ninth flight in September 1956 the aircraft attained a speed of Mach 3.2 and an altitude of 126000 feet, however, it subsequently crashed, killing pilot Captain M. Apt.

395.DOUGLAS X-3 This single seat needle-nosed research aircraft was built to explore sustained high speed, high altitude flight. The plane was extensively instrumented with pressure and heat sensors. The X-3 structure made extensive use of titanium. First flown in late 1952, the aircraft was turned over to the NACA in 1953, under whose auspices a total of 20 research flights were made; the X-3 program was terminated in May of 1956. Powered by two Westinghouse J34 turbojet engines each of 4200 pounds thrust, the 27000 pound X-3 had a wingspan of 22 feet eight inches and a length of 66 feet nine inches, and was designed for a speed of Mach 3.0.

N396.ORTHROP X-4 The single place X-4 research aircraft was designed to investigate the characteristics in subsonic flight of tail-less swept wing airplanes. Two X-4s were built by Northrop, the first being flown in late 1948. The thin wing was highly swept at 40 degrees and was equipped with elevons for both lateral and longitudinal control. The 6900 pound X-4 Bantam was powered by two Westinghouse J30 jet engines each of 1600 pounds thrust, giving a high speed of about Mach 0.9 at altitude. It was a small aircraft with a span of only 26 feet ten inches and even less length. The two X-4s were flown extensively until April 1954 when the test program was terminated. Both aircraft are now in museums.

397.BELL X-5 The X-5 was a single place wing sweep research aircraft built to determine the effect of various degrees of wing sweep angle on performance and handling qualities during flight, and led to the design of later operational swing wing aircraft. The X-5 was based on the German Messerschmitt P1101 of World War II. Two X-5s were built. After a first flight in June 1951, the X-5 was operated by both the USAF and NACA. One plane crashed in October 1953 at Edwards Air Force Base. Powered by an Allison J35 turbojet engine of 4900 pounds thrust, the X-5 had a design gross weight of 9900 pounds and a high speed of 650 mph. Wingspan varied from 30 feet ten inches to 18 feet seven inches as the wing sweep changed.

398.BELL X-5 A flight view of the Bell X-5 shows the wing in the slower speed minimum sweep position. The wing sweep angle could be varied in flight between 20 and 60 degrees, with design compensation for change of aircraft center of gravity. Full span leading edge slats were incorporated, as were dive brakes alongside the fuselage.

398-A.Ryan X-13 Among the many aircraft designs for a jet powered vertical flight aircraft was the X-13 Vertijet, a so-called VTOL "tailsitter". The plane was designed to take off and land vertically from a ground service trailer and normally had no landing gear for such flight. After a 1953 USAF contract, first flight with a jury-rig tricycle landing gear was made in December 1955. First transition flight occurred in November 1956. Jet exhaust deflection was used for low speed vertical flight control. In 1957 a complete flight cycle was made. No operational aircraft type came from this test work. Powered by a Rolls Royce Avon jet engine of 10000 pounds thrust, the 7500 pound gross weight delta wing X-13 had a wingspan of 21 feet and length of 24 feet. The photo was taken at the NASM Silver Hill storage facility in February 1979.

399.BELL X-14 A vertical takeoff and landing (VTOL) test bed aircraft to experiment with the concept of diverting jet engine exhaust to achieve vertical takeoff, the first hovering flight of this X-14 took place in early 1957. Compressed air nozzles provided hovering control. Initial transitional flight took place in May 1958. After USAF evaluation the X-14 was turned over to NASA in 1960 and engines were changed to General Electric J85s. The single place aircraft had a maximum forward speed of about 160 mph; wingspan was 34 feet and length 25 feet.

400.NORTH AMERICAN X-15 An X-15 is shown hanging in the National Air and Space Museum in Washington, D.C. The X-15 was designed as a manned research tool to explore problems of aerodynamic heating, stability, and control, and reentry into the atmosphere from extremely high altitude, high speed flight. Three aircraft were built with special materials employed in the design. Rocket flight controls were designed for use outside the atmosphere. Initial flights starting in September 1959 of the first two X-15As were made with two Reaction Motors rocket powerplants each of 8000 pounds thrust; later, one Thiokol throttle-capable rocket motor of 57000 pounds maximum thrust was employed. Flights were made from a B-52 mother plane. Eventually, a speed of over 4500 mph and an altitude of 354000 feet was attained. One X-15 was lost during tests. The program was stopped in late 1968 after 199 X-15 flights had been made.

401.HILLER X-18 The X-18 of the late 1950s was an experimental tilt wing turboprop powered VTOL airplane using the fuselage of a Chase YC-122 transport, upon which was mounted a 90 degree tiltable wing with two Allison T40 turboprop engines each of 5850 equivalent shaft horsepower, and driving six blade 16 foot diameter Curtiss Turbolectric dual rotation propellers. A Westinghouse J34 jet engine provided divertable jet thrust at the tail for low speed and hover control. First flight as a conventional aircraft took place in November 1959. The wing was tilted up 50 degrees, but flight was described as "hairy" by the test pilot. In 1961 the program was resricted to using the aircraft in a tie-down rig for propeller downwash tests. Gross weight was 33000 pounds, and conventional flight high speed was 250 mph.

402.CURTISS X-19 A six to eight place turboprop VTOL aircraft making a first hover flight in June 1964. Two aircraft had been ordered by the USAF as Tri-Service test vehicles, and the first rolled out in the summer of 1963. A tandem fixed wing four tilting propeller aircraft, the X-19 was powered by two fuselage-mounted Lycoming shaft turbine engines each of 2200 horsepower driving propellers through a mechanical transmission system. The aircraft had a design gross weight of 12300 to 13600 pounds, and a transmission-limited high speed estimated at 440 mph at 15000 feet. The first X-19 crashed in mid-1965 after losing a propeller part way through transition flight near Atlantic City, New Jersey. The second aircraft was only partially completed before the program was canceled.

403.MCDONNELL XV-1 A convertiplane in which the rotor used for vertical takeoff could be "unloaded" by the fixed wing in forward flight, the 4800 pound XV-1 used a Pratt and Whitney 450 horsepower piston engine to drive compressors providing air to rotor tip jet units. Fuel was also fed to these jets. Small anti-torque rotors were used at the tail. A pusher propeller provided thrust for forward flight. The aircraft was used in a joint program of the Army and USAF with first flight including conversion in April of 1955. Two aircraft were built, and one reached 200 mph in August 1956. A year later the program was discontinued.

404.BELL XV-3 Developed using another joint Army and USAF program, cross-shafted tilt rotors providing thrust throughout the regime from vertical to horizontal flight. Electric motors were used to tilt the rotors; the wing remained fixed to the fuselage. Two aircraft were built. First vertical flight took place in 1955, but the aircraft crashed in October 1956. Using the second 4800 pound XV-3, a full conversion from vertical to horizontal flight was made in 1958 using a 450 horsepower Pratt and Whitney engine, and many more conversions were completed later. Rotor problems were encountered in forward flight, however, and a completely new rotor design was installed for more tests in the mid-1960s. The XV-3 was a useful tilt rotor test vehicle with well over 100 flying hours, and led to further development of tilt rotor VTOL aircraft, such as the XV-15 and V-22.

405.LOCKHEED XV-4A A twin jet-engined VTOL research vehicle, the XV-4A Hummingbird project was funded by the Army with two aircraft built. Vertical lift capability was effected by jet engine efflux entraining additional free air using an ejector effect. Jet nozzles at the extremities provided hover control. Diverter valves provided for jet thrust in forward flight. Power for the 7200 pound gross weight XV-4A was provided by two modified Pratt and Whitney JT12 jet engines and a forward flight high speed of over 500 mph was expected. In November 1963 a first transition from vertical to horizontal flight was accomplished, but a few months later the first Hummingbird crashed. The second aircraft was revised to use direct jet lift, since the ejector was deemed unsuccessful. In 1969 the second aircraft, an XV-4B, crashed, and the program was ended.

V406.ERTOL VZ-2 A joint Army-Navy funded program was employed to develop the VZ-2, a flying test bed used to explore the concept of a tilt wing VTOL aircraft. A single Lycoming shaft turbine engine of 860 horsepower drove two rotor/propellers on a tiltable wing through an interconnected mechanical drive system. Ducted fans in the tail assembly provided hover control in pitch and yaw, with main rotors controlling roll in hover. First flight took place in August of 1957 with a transition flight effected almost a year later. The VZ-2 successfully proved the tilt wing concept with 448 flights and 34 full conversions through the flight regime by the middle of 1961, at which time NASA performed additional testing.

407.RYAN VZ-3 The VZ-3 was a small V/STOL research aircraft constructed for the Army to investigate obtaining vertical lift via the deflected propeller slipstream principle with a complex double flap system performing the deflecting and wing end-plates confining the flow. A jet deflection device at the tail directed engine exhaust for low speed pitch and yaw control. Conventional controls were used otherwise. Testing started in February 1958, and a first flight took place in December 1958. Very short takeoffs (STO) and some hovers were accomplished, along with transitions to forward flight. Later the aircraft suffered a crash during tests. Powered by a single Lycoming T53 turboshaft engine of 1000 horsepower driving two interconnected propellers, the single place VZ-3 weighed 2600 pounds loaded and had a wingspan of 23 feet five inches and length of 27 feet eight inches.

408.DOAK VZ-4 Another research aircraft built for the Army to investigate V/STOL flight concepts, the VZ-4 was a tilt ducted fan machine with the shrouded fans at the wingtips tiltable approximately 90 degrees to achieve vertical flight. The single 840 horsepower T53 turboshaft engine in the fuselage drove the fans through mechanical shafting in the wings, and jet exhaust went to a tail device to be directed for low speed and hover control in pitch and yaw. Low speed roll control was gained using the wingtip fans. First flight occurred in February of 1958, and after tests at Edwards AFB the Army accepted the aircraft in September 1959. The Doak was a two-place aircraft with a high speed of 230 mph. Gross weight was 3200 pounds and overall wingspan was 25 feet six inches.

409.VOUGHT-HILLER-RYAN XC-142A In early 1962 a project go-ahead was given on a Tri-Service basis for design and construction of an operational cargo-type four turboprop engined tilt wing V/STOL aircraft known as the XC-142A in a five aircraft program.. A first conventional flight was made in August 1964, and a first hover flight in November of that year. Propeller thrust in hover was less than expected and empty weight greater than anticipated, however, all five aircraft were delivered in 1965 and a thorough military evaluation test program involving 353 flight hours was completed in December of 1965. The XC-142A had a wingspan of 67 feet six inches, a 30 foot long fuselage cargo compartment, a maximum design gross weight of 43700 pounds, and four intershafted General Electric T64 engines each of 2805 takeoff horsepower, plus 310 pounds of jet thrust, which drove four 15.625 foot Hamilton propellers. Maximum speed anticipated was 430 mph at medium altitude. With the advent of the Vietnam War and the prevailing Department of Defense lack of confidence in V/STOL concepts, the XC-142A program was canceled in the later 1960s.

410.BELL XV-15 The XV-15 was a twin engine tilt rotor V/STOL research aircraft with a crew of two and cabin space for nine people. Two 25 foot diameter three blade cross-shafted rotors were driven by Lycoming turboshaft engines, each of 1550 shaft horsepower. Wingspan was 35 feet two inches and design and maximum takeoff gross weights were 13000 and 15000 pounds respectively. Estimated maximum speed was 382 mph at 17000 feet. Both engines and rotors tilted 90 degrees in wingtip nacelles. Two aircraft were built under auspices of NASA and the U.S. Army to explore limits of the tilt rotor flight envelope and assess application to military and civil needs. First hover flight took place in May 1977. Wind tunnel tests were completed in 1978. Extensive flight testing was accomplished on these aircraft, leading to the later development of the Bell Boeing V-22 Osprey tilt rotor aircraft.

411.BELL-BOEING V-22 The V-22 Osprey is a twin-engined tilt rotor cargo V/STOL aircraft of approximately 25 tons gross weight currently under engineering and manufacturing development. The Marines and Navy have the largest requirement for the aircraft, however, 50 CV-22B versions are needed for the U.S. Special Operations Command (SOCOM). At least four prototypes were built, and a first flight of #1 aircraft took place in March of 1989; construction of four additional production-representative aircraft began in early 1994, with a first flight anticipated in December 1996 and operational evaluation in 1999. IOC for the SOCOM version would be in 2005. Early in 1996 a total of over 1076 hours have been accumulated in over 940 flights with V-22s. Two Allison T406 shaft turbine engines each of 6150 shaft horsepower drive two 38 foot diameter rotors in tilting nacelles at the wingtips. The aircraft has a crew of two and accomodation for 24 troops or 12 litters. Maximum level speed is about 315 mph.

TRAINER

Shortly after World War II in 1948 revisions were again made to the system designating Army and Air Force airplanes. In the trainer plane area, instead of using the earlier PT-/BT-/AT- system for designating primary, basic, and advanced types, the simple T- for trainer was put in place. This system was applied regardless of the trainer type involved. Planes still in use at that time were redesignated—for instance, the AT-6 Texan advanced trainer became just the T-6, and the Boeing Stearman PT-17 became the T-17.

The new trainer series started off with the North American T-28A Trojan of 1949, apparently because the highest trainer number earlier was the PT-27. The T-28A incorporated the tricycle landing gear now on most all U.S. combat types. The T-28A had approximately the power and performance of a just pre-World War II pursuit aircraft, and was used as a basic trainer in the 1950s since it was a pretty "hot" airplane for a brand new pilot trainee to step into for his initial flying.

The T-29 trainer of 1949 and several years afterwards was a military version of the popular postwar Convair 240 twin-engined Double Wasp powered airliner, showing the wide variety of aircraft included now under the new T- designation system. Built in several versions, the T-29 was variously a trainer for navigators and radar operators.

The two seat "T-bird" trainer came into being in early 1948 and was originally designated a TF-80C, a jet trainer version of the F-80C Shooting Star fighter by Lockheed. Shortly, this aircraft was redesignated in accordance with the new trainer system as a T-33A. In the next twenty years or so the T-33A T-bird became one of the most well known aircraft in the U.S. military inventory and also world-wide.

In 1953, after a competition with the Temco YT-35, Beech won a USAF contract to supply a 225 horsepower piston-engined two place low wing monoplane variant of their Model 45 Bonanza to the Air Force for use as a primary trainer. Almost the last of the piston-engined trainers, the T-34 was pretty much alone in the USAF. It had a long life (well into the 1970s), and it was finally supplanted by the Cessna T-37 twin jet airplane.

The Cessna T-37 was the first jet-engined primary trainer, originally envisioned as a basic trainer, but proving itself for a while equal to the primary training task. Interestingly, the T-37 reverted to a side-by-side seating arrangement for instruction, the first in a primary trainer since the early 1920s. The T-37A first flew in late 1955, and a later T-37B updated version came out in 1959. All the T-37As went through a modernization program begun in the early 1960s. The T-37C is similar to the T-37B but provides for armament and wingtip fuel tanks, and has been supplied to several foreign countries under the Military Assistance Program. The replacement for the T-37 was to have been a new jet trainer built by Fairchild Republic, however, the program encountered great difficulties and was canceled by the USAF, leaving the T-37 as the Air Force primary training airplane. Republic then went out of business.

The Northrop T-38 is an unusual trainer because of its performance capabilities as a twin-engined supersonic jet airplane, again illustrating the diversity of aircraft lumped under the T- for trainer designation. First flown in 1960, it started service early the next year. The crew sits in tandem on rocket powered ejection seats.; the two afterburning engines are equipped with variable exit exhaust nozzles, and the aircraft is in most respects like a fighter, and except for armament is indeed much like the Northrop F-5B fighter.

Another twin-engined jet trainer of a different sort was entered as a private venture by North American in the UTX (Utility Transport Experimental) competition of the mid-1960s. The company hoped to not only win the USAF trainer competition, but also to have a design in hand for a corporate jet as well. Their effort was successful, and a production T-39A was delivered to the USAF in 1960 as a six place transport with many more following. The aircraft was used primarily as a utility transport.

Again the USAF had second thoughts of putting raw new pilot candidates directly into a T-37 twin jet airplane as an initial trainer. So, in the early 1960s they ordered a high wing two to four place Cessna 172 monoplane to provide some initial training in a 145 horsepower piston-engined aircraft. This was the T-41 Mescalero with a high speed of 138 mph. Many other countries also ordered the Mescalero.

The Army T-42 Cochise was a civil Beech Baron ordered as a light transport and instrument trainer in the mid-1960s. The T-42 was also employed as an Army VIP transport and could carry pilot, copilot, and two passengers with excellent accommodations.

The Boeing T-43A is a military navigation trainer variant of the 737-253 twin-engined civil transport, a replacement for the Convair T-29 navigation trainer. The T-43A first flew in USAF colors in 1973 with 19 aircraft ordered. The cabin has accommodations for three instructors, 12 trainees, and four advanced trainees. Installed equipment includes the Undergraduate Training System using Honeywell T-45 electronic simulators. Other items included are celestial, radar, and inertial navigation systems, LORAN, and other radio systems. The plane is powered by two Pratt and Whitney turbofan engines.

373.NORTH AMERICAN T-28A In 1948 all trainers were redesignated under a T designator, and the T-28A Trojan was the first new airplane so designated, continuing the old PT- designator number, the last of which was a Stearman PT-27 primary trainer biplane. The winner of a 1948 competition to replace AT-6/T-6 Texan trainers, the prototype flew first in late 1949. Distinguished by a tricycle landing gear, the T-28A was built in quantity by North American with a total of 1194 produced in the early 1950s.

374.NORTH AMERICAN T-28A The T-28A Trojan was a more powerful and heavier trainer than the T-6 with a Wright R-1300 800 horsepower radial engine and a gross weight of 6360 pounds, a 200 horsepower and 1100 pound margin over the earlier Texan. Span was 40 feet, while high speed was about 280 mph, significantly higher than that of the T-6. It was quite complex for a primary trainer, and was more properly a basic or advanced trainer.

375.Convair T-29B The big twin-engined Convair T-29 trainer was an Army version of the Convair 240 civil transport aircraft, and was produced in several military sub-variants. Originally designated as an advanced trainer, the XT-29 first flew in 1949. A total of 48 non-pressurized T-29A versions were delivered in 1950 as navigation trainers. Two years later, a T-29B navigation and radar operator trainer with a pressurized fuselage appeared, with Convair producing 105 aircraft. The T-29B was powered by two Pratt and Whitney R-2800 Double Wasp engines each of 2400 horsepower giving a high speed of about 295 mph. Wingspan was 91 feet nine inches, and normal gross weight was 43600 pounds.

376.CONVAIR T-29C Another variant of the T-29, the C version was a navigation trainer like the B model and was introduced in 1953. Convair built 119 T-29C trainers with water injectable Double Wasp engines uprated to 2500 horsepower and other minor modifications. The T-29s could also be converted into personnel transports.

377.Convair T-29D Still another version of the twin engine Convair, the Army T-29D of 1953 was equipped to train bombardiers and navigators using the latest equipment installed in USAF bombers of the period. Convair put out 93 T-29D aircraft using two 2500 horsepower Double Wasp engines. Weights and performance were similar to earlier models.

378.LOCKHEED T-33A Never was a single seat Army/Air Force fighter aircraft so successfully adapted into a two seat trainer as in the case of the Lockheed F-80 Shooting Star turning into the T-33A "T-Bird". Extension of the F-80 fuselage slightly over three feet allowed a second cockpit. Originally labeled a TF-80C and first flying in 1948, the aircraft became a T-33A the next year. In the next decade the T-Bird became very well known with over 5600 being produced by Lockheed. The T-33A served USAF, Air Force Reserve, and Air National Guard units well into the 1970s. The T-Bird shown belonged to the 182nd Fighter Interceptor Squadron in Texas; the photo was taken in January of 1972. Provision was made in the nose for installation of two .50 caliber machine guns.

379.LOCKHEED T-33A The clean lines of the T-33A are shown in this flight photo of an early USAF jet flight demonstration team. Notable is the absence of the usual wingtip fuel tanks. The T-33A used the same engine as the F-80C, a 4600 pound thrust Allison J-33 centrifugal flow jet based on the British Rolls Royce Nene. With a gross weight of about six tons, the T-bird had a maximum speed of about 540 mph at high altitude and had a ceiling of over 47000 feet.

380.LOCKHEED T-33A Many T-Birds were used in the air services of well over 30 different countries. The aircraft, shown in August 1973, is one of over 180 produced under license in Japan by Kawasaki for the Japanese Air Self Defense Force. A single piece canopy was used to cover both cockpits; it is shown in the fully raised position. Wingtip fuel tanks with an aft stabilizing fin outboard were almost always carried on the T-Bird.

381.BEECH YT-34 The T-34 Mentor was the final USAF piston-engined primary trainer airplane. The aircraft shown is one of three service test Beech Model 45 airplanes evaluated by the USAF as YT-34s in 1950. After extensive testing the T-34A was put into production as the USAF primary trainer of the 1950s. The two place tandem seat low wing Mentor with retractable tricycle landing gear was powered by a 225 horsepower Continental engine, giving a high speed of about 185 mph. The T-34A had a loaded weight of 2900 pounds and a wingspan of 32 feet, ten inches.

382.CESSNA T-37B The T-37 was the first USAF jet primary trainer, making that service truly an all jet air force. The two place side-by-side seating twin engine trainer flew first in late 1954 and a few were ordered for service test. In the late 1950s the production T-37A was utilized as a primary trainer instead of the piston-engined T-34. A total of over 400 A models were produced, and then a B version came into service with uprated engines. Powered by two Continental J69 jet engines each of just over 1000 pounds of thrust, the T-37B had a high speed of about 400 mph at high altitude. It had a wingspan of 33 feet nine inches and a gross weight of about 6400 pounds. In the 1960s the T-37 trainer was developed into an AT-37D armed light counter-insurgency aircraft with General Electric J85 engines.

383.NORTHROP T-38A The T-38A Talon was based on a Northrop design for an inexpensive lightweight fighter, the N-156. The design evolved into a tandem two seat trainer as well as the single seat F-5 fighter. The T-38 trainer version was the first supersonic aircraft of the type. In 1958 Northrop received a contract for seven YT-38 service test airplanes, which started flying the next year. Successful service tests resulted in eventual production, starting in 1960, of well over 1100 Talon trainers for the USAF, along with 46 for West Germany. NASA received two dozen for use in the astronaut program as well, and five were supplied to the U.S. Navy.

384.NORTHROP T-38A The Northrop Talon was an unusual trainer in its ability to achieve a supersonic speed of about Mach 1.25. Powered by twin General Electric J85 afterburner-equipped jet engines each of 3850 pounds maximum afterburner thrust, the six ton gross weight T-38A Talon could reach a speed of over 800 mph at 36000 feet altitude. Initial maximum climb rate at sea level was 30000 feet per minute and the ceiling with half fuel burned off was near 50000 feet. This kind of performance was pretty special for a trainer aircraft!

385.NORTH AMERICAN T-39A A military version of the company's civil Sabre or Sabreliner 40, the six place T-39 was ordered initially in early 1959 by the USAF in a civil prototype evaluation. Production deliveries commenced the next year. Orders in the early 1960s resulted in a total production of 143 T-39As for the USAF and six T-39B versions. It was used as a proficiency trainer and staff transport. Powered by two aft-podded Pratt and Whitney J60 jet engines each of 3300 pounds thrust, the T-39A weighed about nine tons loaded and high speed was about 570 mph at 22000 feet. Navy versions were the T-39D and CT-39E.

386.CESSNA T-41B The Cessna T-41 Mescalero was the militarized version of the company Model 172 four place cabin monoplane. The T-41A was first ordered by the USAF in July, 1964 to provide student pilots with a short period of basic training before moving on to the T-37B jet, and 208 were built. Thirty-four T-41As were supplied to South American governments. The U.S. Army version shown in the photo was a more powerful civil R172E model of which 255 were ordered in August, 1966. Later in 1967 the USAF ordered 45 T-41C versions with fixed pitch propellers specifically for cadet training at the USAF Academy. A further 121 T-41D versions went to various operators, including at least 30 to Columbia.

387.NEG1135 BEECH T-42A The Beech T-42A Cochise was an Army version of the civil Beech Baron B55 four to six place light transport. In February, 1965 the Army selected the type as an instrument trainer, and 65 were produced for that service. In 1971 five T-42As were delivered to Turkey and others went to Peru. The 5100 pound T-42A had two Continental piston engines of 260 horsepower each giving a high speed of about 235 mph and a service ceiling of about 20000 feet. The T-42A shown was photographed in September of 1979.

UTILITY

In 1962 the Army and Air Force initiated a U- for utility aircraft category. Previously, the services often made obsolete former front line service aircraft into utility hacks on an informal basis.

With institution of the utility category several liaison aircraft were redesignated as utility types. As noted in the section on liaison airplanes, the Cessna L-27 became a U-3, the Aero Commander L-26 became the U-4 and U-9, the DeHavilland L-20 Beaver became a U-6, the Piper L-21 a U-7, the Beech L-23 a U-8, the Helio L-28 a U-10, and the Grumman SA-16 Albatross became a U-16. The U-11A was a Navy Piper Aztec E.

The first utility airplane under the new system was a DeHavilland U-1A Otter, a larger version of the earlier Beaver. A converted Canadian "bush" airplane, the Otter was a true utility aircraft—a jack of all trades.

The Lockheed U-2 was placed in the utility category in an effort to obscure its true primary mission of very high altitude strategic reconnaissance over the territory of a potential enemy. Designed and built at Kelly Johnson's Lockheed "Skunk Works" in the late 1950s, the U-2 models had the high aspect ratio wing and low wing loading of a very high altitude aircraft, along with bicycle and wingtip landing gear.

Another aircraft in the utility category was the U-17A/U-17B, a modified Cessna civil Model 185 Skywagon supplied to the USAF for delivery to friendly countries overseas under the U.S. Military Assistance Program.

The U-21 was the Army's version of the civil Beech King Air turboprop airplane used for light transport and miscellaneous duties. The RU-21J is a version of the later Super King Air 200.

The QU-22B was a civil Beech A36 Bonanza revised specially for the USAF Pave Eagle program operated from Thailand during the Vietnam War. This was a special infiltration surveillance program involving aircraft relay to a friendly ground station of data transmitted by acoustic sensors air-dropped over trails used by the enemy.

The AU-23A was a Fairchild Industries/Pilatus light counter-insurgency aircraft version of the Pilatus civil Porter called the Peacemaker, which was lightly armed and could also be used for reconnaissance, leaflet dropping, and loudspeaker broadcasting. Fifteen aircraft were ordered by the USAF in 1970 under the Credible Chase program; thirteen of these were supplied to Thailand.

The Helio AU-24A was a similarly armed version of that company's civil H-550 Stallion, of which 15 were acquired by the USAF in the early 1970s with all but one supplied to Cambodia.

388.DEHAVILLAND U-1A The U-1 Otter was the first aircraft in the new U.S. Army Utility category, and was a militarized version of the DeHavilland Canada civil DHC-3 light transport. Named the Otter, this aircraft was a larger follow-on to the earlier DH Canada Beaver. The U.S. Army bought six YU-1 service test aircraft in 1955. The Otter's success resulted in the purchase of 176 U-1A aircraft. Powered by a Pratt and Whitney R-1340 Wasp engine of 600 horsepower, the four ton gross weight U-1A had a top speed of 160 mph and a service ceiling of 18800 feet. The aircraft could carry a pilot and copilot and up to ten cabin passengers, or alternately six stretchers and four passengers.

389.LOCKHEED U-2 The USAF U-2 high altitude reconnaissance aircraft was exposed to worldwide publicity after the May 1960 shoot-down by the Russians of pilot Gary Powers in his U-2. The U-2 was designated as a utility type to disguise its primary mission; it was flown first in the middle 1950s. Only several dozen aircraft were built, and these were employed through the 1960s in strategic reconnaissance and weather information flights. Normally it was a single place aircraft, but a few were modified as two place U-2Ds. Powered by a special Pratt and Whitney J57 jet engine, the eight ton gross weight U-2 could fly at a maximum speed of about 500 mph. It had a service ceiling of at least 70000 feet and a maximum endurance of approximately eight hours.

390.BEECH RU-21F The Army U-21 was a military version of the Beech King A100 civil transport aircraft and became the first Army airplane with a pressurized cabin. Five U-21F twin turboprop airplanes were delivered to the Army in late 1971 having a crew of two and cabin space for from six to 13 passengers. Powered by two Pratt and Whitney PT6 turboprop engines of 680 equivalent shaft horsepower each, the 11500 pound gross weight U-21 had a maximum cruising speed of 285 mph at 10000 feet, a maximum climb rate of 1960 feet per minute at sea level, and a service ceiling of about 25000 feet.

GALLERY

A-0. Douglas XA-2 A 1926 O-2 conversion with eight guns.

A-1. Curtiss XA-4 A 1928 A-3 conversion to Wasp radial engine

A-2. Fokker XA-7 Loser to XA-8 in 1930 attack competition.

A-2A. Curtiss A-8 One of 13 service test aircraft in 1933.

A-2B. Curtiss YA-10 A 1933 conversion of an A-8 to Hornet engine

A-3A. Consolidated A-11 A 1933 attack version of Y1P-25 pursuit.

A-3B. Consolidated A-11   One of four aircraft with Conquerors.

A-3C. Curtiss A-12   Cyclone powered Shrike attack of 1934.

A-3D. Northrop 2-E Military Gamma demonstrator attack plane.

A-4. Curtiss XA-14 A 1936 predecessor of A-18, Wright radials.

A-5. Northrop XA-16 YA-13 re-engined from Cyclone to Twin Wasp.

A-5A. Northrop XA-16 1935 attack prototype, predecessor to A-17

A-6. Vultee YA-19 The V-11 for U.S. in 1939 with Twin Wasp.

-6A. ultee YA-19 One of five aircraft. Not used in operations

A-7. Vultee V-12 Improved V-11 with more power. Export only.

A-8. Stearman X-100 Initial attack competition entry in 1940.

A-9. Stearman XA-21 MFR X-100 modified with conventional cockpit.

-9A. Stearman XA-21 MFR Loser in attack competition; one only.

A-10. Martin XA-22 Loser in Army competition, winner in exports

A--10A. Curtiss A-25 Army Helldiver, this one for Australia.

A-11. No.American NA-44 **Very similar to A-27s from Siam order.**

11A. Martin A-30A **The Model 187 Baltimore used by Britain.**

A-11B. Vultee A-31 Early version of Vengeance dive bomber.

A-12.Brewster XA-32 One of two prototypes. One seat, four 20mm.

BA-13. eech XA-38   Two prototypes in 1945; 75mm nose cannon.

A-13A. Beech XA-38   Powered by two Double Wasp engines.

A-13B. Fleetwings XA-39 **Mockup of big Fleetwings attack plane.**

A-13C. Curtiss XA-40 **Model of single seat heavy attack plane.**

A-13D. Convair XA-41 One prototype with Wasp Major engine.

A-13E. Northrop A-9A Twin engine loser in fly-off with A-10A.

A-13F. Boeing AT-3 Lower powered PW-9 for advanced training.

A-14. Curtiss AT-4 One of 35 Hawk trainers; later to P-1D.

A-15. Beech AT-10   Over 2000 built of plywood in 1942.

A-15A. Beech AT-11   Bombardier trainer version of civil Model 18

A-16Seversky 2-PA One of two in Sweden. Others became AT-12.

A-16A. Fairchild XAT-13 One prototype bomber crew trainer in 1942.

A-16B. Fairchild XAT-14   Like XAT-13 but with Ranger engines.

A-17. Boeing XAT-15   One of two plywood trainers in 1942.

A-17A. Boeing XAT-15 Trainer built at Wichita, Kansas, only two.

A-41. No. Amer. BC-2 One of three in 1938, developed from BC-1.

A-42. Vultee BC-3 **Competition to BC-1, one only in 1940.**

A-43. Curtiss CW-23 **St. Louis Div. private venture, one only.**

A-44. Douglas BT-2B Basic trainer variant of O-32 observation.

A-44A. Stearman YBT-3 Single PT-9 converted to Wright Whirlwind.

A-44B. Curtiss XBT-4 A 1932 conversion of an O-1E to trainer.

A-44C. Stearman XBT-5 A PT-9 redesignated to basic trainer in 1932

A-44D.Seversky BT-8 One of 30 basic trainers; first low wing.

A-45. No.American BT-9 Whirlwind powered basic trainer.

A-46. No.Amer. BT-9B Major production version; 117 built in 1937

A-47. No.Amer. BT-9C The 1939 model of well-liked trainer.

A-48. No. Amer.Y1BT-10 One BT-9C converted to Wasp engine.

A-48A. Fleetwings BT-12 Stainless steel trainer; one of 24 in 1942.

A-48B. Vultee BT-15 Valiant trainer of 1941 with Wasp Jr. engine

A-48C. Stearman XBT-17 Single prototype of 1942. No production

A-17B. Witteman XNBL-1 Barling long distance night bomber of 1923.

A-18. Curtiss XNBS-4 One of two short distance night bomber types

A-18A. Curtiss B-2 One of twelve Condor bombers of 1929.

A-18B. Keystone B-3A Three dozen converted in 1930 from LB-10As.

A-18C. Keystone LB-7 A 1930 version of the Keystone bomber.

A-18D. Keystone B-4A Hornet B powered bomber of 1932; 12 built.

A-18E. Keystone B-5A Another five place light bomber of 1930.

A-18F. Fokker YB-8 The second XO-27 converted as a bomber.

A-19. Martin B-10B Production version of famous Martin bomber.

A-20. Martin B-12 Martin changed from Hornets to Cyclones.

A-21. Martin YB-12A B-12 with extra fuel runs up in winter.

A-22.Martin B-12A The B-10 and B-12 ushered in modern craft.

A-23. Martin XB-14 One B-12 with Twin Wasp engines.

A-24. Boeing B-17B Early B-17 of 1939; Cyclones with turbos.

A-24A. Boeing B-17B The first "production" B-17; 39 built.

A-24B. Boeing B-17E First fortress with tail gunner in 1942.

A-24C. Boeing B-17H B-17G converted to air-sea lifeboat rescue.

A-25. No.American XB-21 Loser to B-18 in 1937 competition.

A-25A. No.American XB-21 Fine view of Dragon with Two twin Hornets.

A-26. Douglas B-23 One of 38 with Double Cyclones. To UC-67.

A-27. Douglas B-23 New 1939 design; first with tail gun.

A-28. No.American B-25H Mitchell with gun nose and gun side packs.

A-29. Lockheed B-34 A Vega R-37 export version of B-34 of 1942.

A-30. Northrop XB-35 Flying wing heavy bomber prototype of 1946.

A-30A. Boeing XB-38 AAF Fortress Allison engine conversion in 1943.

A-30B. Boeing XB-39 B-29 conversion to 24 cylinder Allisons.

A-31. Boeing YB-40 Heavily armed B-17 as escort; not successful

A-32. Douglas XB-42 Post-war piston bomber, buried Allisons.

A-32A. Douglas XB-42A   XB-42 Mixmaster with jet engines added.

A-38. Douglas XB-43   First U.S. jet powered bomber; two jet engines

A-33A. Douglas XB-43   Two place bomber prototype of 1946

A-34. Boeing XB-44   One of three B-29s with Wasp Major engines.

A-35. Convair XB-46 A 1946 jet bomber losing out to XB-45.

A-36. Boeing B-47B First large production version of Stratojet.

A-37. MARTIN XB-48 One of two straight wing six jet bombers.

A-37A. Martin XB-48 Bomber had bicycle landing gear, outriggers.

A-38. Northrop YB-49 Jet engine version of XB-35 flying wing type

A-38A. Northrop YRB-49A Reconnaissance version of jet flying wing.

A-38B. Martin XB-51 Three jet swept wing medium bomber. B-57 won

A-38C. Martin XB-51 Configuration of the XB-51 was unusual.

A-38D. Martin XB-51 Bicycle gear and a variable incidence wing.

A-39. Boeing B-52D Showing the bicycle gear and big wing tanks.

A-39A. Convair YB-60 Swept wing eight jet version of B-36.

A-40. Rockwell B-1A Swing wing four turbofan bomber. Canceled.

A-40A. Rockwell B-1A B-1A takes off with wing extended.

A-40B. No. American XB-70 Mach 3.0 Valkyrie at USAF Museum.

A-48D. Douglas C-1/Fokker C-2A C-1 refuels C-2A in early attempt.

FA-49. ord C-4A One of four 5-AT trimotors procured in 1931.

A-50. Fairchild C-8   Eight civil 71 types converted from F-1s.

A-50A. Ford C-9   A Ford C-3A re-engined with Whirlwinds.

A-51. Lockheed Y1C-12 A single civil DL-1 Vega type bought 1931.

FA-52. okker C-14A One of 20 civil F-14 types used in 1931.

A-52A. Lockheed Y1C-17 One Speed Vega in 1931 with Wasp engine.

A-52B. Lockheed Y1C-23 One commercial Altair 8-D bought in 1932.

A-52C. American Y1C-24 A militarized civil Pilgrim of 1932.

A-58. Bellanca C-27B A 1934 Army version of the civil Airbus.

A-54. Bellanca C-27C A 1934 Airbus with uprated Cyclone engine.

A--55. Sikorsky C-28 A 1932 Army version of civil S-39.

A-55A. Douglas C-29 Two Dolphins revised from C-26B with Wasps.

A-55B.Curtiss YC-30 Two civil T-32 Condors bought in 1934.

A-56. Douglas C-33 Militarized DC-2 with Cyclone engines.

A-57. Douglas C-34 One of two more DC-2s procured in 1936.

A-58. Lockheed Y1C-36 A Model 10A Electra bought in 1937.

A-58A, Lockheed Y1C-37 Another 10A Elecra of 1937, ten place.

A-59. Lockheed C-40B Model 12A Electra Jr., tricycle gear.

A-59A.Curtiss C-46A First mass produced Commando version.

A-60. Curtiss C-46E St. Louis Div. Commando ; 17 bought 1943.

A-61. Douglas C-49 A civil DC-3 drafted by the Army

A-62. Douglas C-49 All had Cyclone engines, carried 21 people.

A-63. Douglas C-53 Skytrooper DC-3 type; 193 bought in 1941.

A-64. Douglas C-54G One of 76 Skymasters purchased in 1945.

A-64A. Curtiss XC-55 Civil CW-20 of 1940. Scrapped in England.

A-64B. Lockheed C-60A Civil Model 18 Lodestar of 1942.

A-64C. Boeing C-73 One of 27 247Ds drafted in 1942.

A-64D. Douglas C-74 A 1946 Globemaster with Wasp Major engines.

BA-64E.oeing C-75 One of the five 307s drafted in 1942.

A-64F. Curtiss C-76 A few wooden Caravans were made in 1942-3.

A-64G. Fairchild C-82A Twin boom Packet cargo plane of 1946.

A-64H. Boeing XC-108 B-17E Fortress revised as 1943 transport.

A-64I. Curtiss XC-113 XC-46C Commando as turboprop test bed.

A-64J. Douglas C-118A **Civil DC-6 converted for military use.**

A-65. Fairchild C-119C **Second production version of flying boxcar.**

A-65A. Fairchild XC-120 Pack-plane test version of C-119.

SA-65B. troukoff YC-122C Avitruck post-war assault transport.

A-66. Chase XC-123A First jet transport with B-47 nacelles.

A-67. Fairchild C-123B Early Provider version, Stroukoff design.

A-68. Douglas C-124A Globemaster II, 204 aircraft early 1950s.

A-68A. Northrop YC-125B Service test Raider trimotor, few only.

A-68B. Cessna LC-126C Army version of civil Cessna 195.

A-68C. Stroukoff C-134A Pantobase version of C-123 assault transport

A-69. LTV XC-142A Tri-service tilt wing V/STOL, five only.

A-70. Fairchild F-1 Army photo plane based on civil F-71.

A-71. Lockheed F-5B Photo Lightning of 1943, converted P-38G.

A-72. Hughes XF-11 One of two 1946 reconnaissance planes.

A-72A. Hughes XF-11 One had dual rotation propellers, crashed.

RA-73. epublic XF-12 The single Rainbow prototype, Wasp Majors.

A-74. Republic XF-12 The XF-12 was overtaken by jet types.

A-75. Northrop F-15 Photo version of XP-61E in 1946.

A-104B. Curtiss 18B   Two two-place biplanes of 1919, Kirkham eng.

A-104C. Loening PW-2   Monoplane pursuit of 1921; two built.

A-104E. Curtiss PW-8A 1924 prototype with skin cooling for D-12.

A-104D. Gallaudet PW-4 Single place 1922 pursuit; never flew.

A-104F. Curtiss XPW-8A   One of three converted for racing in 1924.

A-104G.Boeing XPW-9   Third of three XPW-9 prototypes n 1923.

A-105. Curtiss P-1A **Curtiss Hawk of 1926 with D-12 engine.**

A-105A. Curtiss P-1B **1927 improved version of Curtiss Hawk.**

A-106. Curtiss P-2 One of five 1926 Hawks with engine change.

A-107.Curtiss XP-3A Wasp powered Hawk tested with engine cowl.

A-108. Curtiss XP-3A   Cowled Wasp powered Hawk for racing.

A-108A. Boeing XP-4   PW-9 type in 1926 with Packard engine.

A-109. Curtiss P-5 Super-Hawk with turbosupercharged D-12.

A-110. Curtiss P-5 One of five Super-Hawks in 1928.

A-111. Curtiss XP-6 A P-2 airframe with Conqueror as racer.

A-112. Curtiss XP-6A 1927 P-1A Hawk with XPW-8A wings for racing.

A-112A. Curtiss P-6A One of several with Conqueror engine.

A-113.Curtiss P-6B One Conqueror powered Hawk Hoyt Special.

A-113A. Curtiss P-6D A 1931 P-6 conversion to turbosupercharging.

A-113B. Curtiss XP-6E Initial P-6E Hawk with Prestone cooling.

A-113C. Curtiss XP-6F P-6E with turbosupercharger, closed cockpit.

A-113D. Curtiss XP-6G P-6E with revised powerplant.

A-113E. Curtiss XP-6H   A P-6E with new wings and six guns.

A-114. Boeing XP-7   A PW-9D for testing Conqueror engine, 1928.

A-114A. Boeing XP-7   Water cooled Conqueror in PW-9D pursuit.

A-114B. Boeing XP-8   One 1928 prototype with a Packard engine.

A-114C. Boeing XP-9 **A 1930 monoplane pursuit prototype only.**

A-115. Curtiss XP-10 **Single 1930 pursuit prototype, skin cooling.**

A-116. Curtiss XP-10 Conqueror engine, upper wing gulled.

A-116A. Curtiss XP-10 Plywood wings, no production.

A-117. Thomas Morse XP-13 One Viper pursuit of 1930, Curtiss engine.

A-118. Boeing XP-15 A monoplane pursuit of 1929, never purchased

A-118A. Boeing XP-15 Wasp powered prototype; high speed 185 mph.

A-119.Curtiss XP-17 A P-1 converted in 1930; Wright engine test.

A-120. Curtiss YP-20 Once a P-11, then YP-20 in 1931 with Cyclone

A-120A. Curtiss YP-20 Slicked up 1931 pursuit, but P-6E ordered.

A-121. Curtiss YP-20   YP-20 in October, 1930 with short stacks.

A-122. Curtiss XP-22   Another Hawk type; a P-6A modified in 1931.

A-123. Curtiss XP-22 **Ring cowl on Conqueror with special radiator**

A-124. Curtiss XP-23 **The last biplane Hawk in 1932 with turbo.**

A-124A. Lockheed YP-24 Two seat Detroit Lockheed plane of 1932.

A-125. Consolidated Y1P-25 Development of YP-24; two planes in 1933.

A-125A. Boeing XP-26 One of three prototypes for P-26A in 1932.

A-125B. Boeing YP-29 Advanced pursuit of 1934, retractable gear.

A-125C. Boeing YP-29B Modified YP-29, open cockpit, revised wing.

A-126. Consolidated P-30 A 1934 development of the Y1P-25.

A-126A. Curtiss XP-31 Original version of Swift pursuit in 1932.

A-126B. Curtiss XP-31 Swift revision with conqueror ; P-26 rival.

A-126C, Curtiss XP-31 Not so Swift; 25 mph slower than P-26A.

A-127. Curtiss XP-31 Swift with wing leading edge slats open.

A-127A. Vought V-141 Losing competitor in pursuit competition.

A-127B. Seversky AP-1 Company owned prototype for P-35.

A-127C, Seversky P-35 A surviving P-35 at the USAF Museum.

SA-128. eversky EP-1 Export model in Sweden; equivalent to P-35A.

A-129. Seversky AP-9 Advanced design version of P-35. New wing.

A-129A. Curtiss 75 Original 1935 prototype leading to P-36A.

CA-129B. urtiss Y1P-36 One of three service test planes in 1937.

CA-129C, urtiss P-36A A restored aircraft in the USAF Museum.

A-129D. Curtiss P-36A Twin Wasp powered P-36A running up.

A-130. Curtiss H75A Export Hawk to France with Cyclone engine.

A-131. Curtiss YP-37 Allison powered service test pursuit

A-131A. Curtiss XP-37 One turbosupercharged experimental in 1937.

A-131B. Curtiss YP-37 **January 1939 photo of service test type.**

A-132. Lockheed XP-38 **Original XP-38 in 1939, later crashed.**

A-133. Lockheed XP-38 Another view of the Lightning prototype.

A-134. Lockheed P-38H Last version with wing-mounted intercoolers.

A-134A. Lockheed P-38M P-38L revised to a two place night fighter.

A-135. Bell YP-39 Service test Airacobra shows clean lines.

A-136. Bell P-39C   A P-39C with no wing guns is illustrated.

A-137. Bell P-39F   Production Airacobra of 1942; Aeroprop.

A-138. Bell P-39L A 1943 production version with drop tank.

A-139. Bell P-39N Over 2000 built in 1943, Allison engine.

A-140. Curtiss XP-40   Original version of prototype in 1938.

A-141. Curtiss XP-40   Another view of XP-40 with rear radiator.

A-141A. Curtiss P-40 **One of 200 original production versions.**

A-141B. Curtiss P-40Es **New aircraft ready for Buffalo flight test.**

A-142. Curtiss P-40F **First Packard Merlin powered P-40 type.**

A-143. Curtiss P-40K **One of 1300 Allison powered models, 1942.**

A-144. Curtiss P-40L **Another 1943 Merlin powered fighter.**

A-145. Curtiss TP-40N **Two place P-40 proposed as a trainer.**

A-145A. Seversky XP-41   Last P-35, revised gear and powerplant.

A-146. Curtiss XP-42   One of many cowl arrangements tested.

A-147. Seversky AP-4   Company prototype for P-43, aft turbo.

A-148. Republic P-43A   Predecessor in 1941 to later P-47.

A-148A. Curtiss XP-46A Planned P-40 replacement, two in 1942.

A-149. Republic XP-47B Prototype Thunderbolt at plant in 1941.

A-149A. Republic P-47D Bubble canopy Jug at Caldwell, N.J.

A-150. Republic XP-47J High speed model of 1944 with six guns.

A-150A. Lockheed XP-49   P-38 tested with Continentals in 1942.

A-150B. Grumman XP-50   1940 twin engined fighter with Cyclones.

A-151. No.American TP-51D One of few factory two seaters.

A-151A. Convair XP-54 Big pusher with Lycoming engine in 1943.

A-152. Curtiss XP-55 Ascender tail first pusher of 1942.

A-152A. Curtiss XP-55 Allison powered pusher showing arrangement.

A-152B. Northrop XP-56 Modified flying wing fighter of 1943.

A-152C. Lockheed XP-58 Twenty-one ton P-38-like bomber destroyer.

A-152D. Curtiss XP-60 Intended P-40 replacement with Merlin, 1941.

A-152E. Curtiss XP-60A Three 1942 prototypes with Allison engines.

A-153. Curtiss XP-60C An XP-60A with Double Wasp, dual propellers.

A-153A. Curtiss XP-60C Fine flight view of 1943 prototype.

A-153B. Curtiss XP-60D   Modified XP-60 type in January, 1943.

A-153C.Curtiss YP-60E   Final P-60 type, converted P-60A.

A-154. Northrop XP-61 Big night fighter prototype of 1942.

A-154A. Northrop XP-61E Modified P-61B to two place escort fighter.

A-154B. Curtiss XP-62 Large 1943 high altitude prototype.

A-154C. Bell RP-63C One of 200 target planes in 1944.

A-155. Bell P-63F As racer in Reno; two built in 1945.

A-156. No. American P-64 Six planes drafted from Siamese order.

A-156A. No. American P-64 Was NA-50; Wright Cyclone engine.

A-156B. Vultee P-66 Lineup at Karachi, India, October, 1942.

A-156C. Douglas P-70 A-20 converted as a night fighter.

A-156D. Curtiss XP-71 Model of huge fighter never built.

A-156E. Republic XP-72 A P-47 with the Wasp Major engine.

A-156F. Fisher XP-75 One of several; parts from other planes.

A-156G. Bell XP-77 Two wooden light fighters, Ranger engines.

A-156H. Northrop XP-79B 1945 Twin jet with prone pilot position.

A-156I. Lockheed P-80C Final jet Shooting Star version.

A-156J. Lockheed P-80R Special racing version of the P-80.

A-157. Bell XP-83 Two 1945 twin jet prototypes.

A-158. Republic XP-84H Turboprop F-84 with supersonic propeller.

A-159. McDonnell XF-85 Goblin parasite plane for B-36 bomber.

A-160. Northrop F-89D Scorpion night fighter, wingtip rocket pods.

A-160A. Northrop F-89H New Scorpion version with guided rockets.

A-161. Lockheed XF-90 Swept wing prototype, no production.

A-162. Republic XF-91 Jet, rocket powered, inverse taper wing.

A-163. Lockheed F-94A Initial version with radar and gun nose.

A-163A. No. American F-100D Ground attack version of Super Sabre.

A-163B. Convair XF-102 Interceptor prototype before area ruling.

A-164. Republic XF-103 Artist's drawing of turbo-ramjet fighter.

A-164A. Republic F-105B Initial production version of big Thud.

A-164B. Northrop F-20 Tigershark.

A-165. Bell YFM-1 Multiplace fighter at New York World's Fair.

A-166. Bell YFM-1 Twin nacelle pusher Airacuda, no cannons.

A-76. Aeronca L-3C Grasshopper type; 490 delivered in 1943.

A-77. Ryan L-17B Navion procured by Army; 158 aircraft.

A-78. Cessna XL-19B A 1952 turboprop with Boeing T-50 engine.

A-79. Beech L-23B **One of 40 liaison Seminoles; later utility.**

A-79A. Beech L-23F **A later Seminole version, changed to U-8F.**

A-80. Cessna L-27B **From civil Model 310; later a U-3B.**

A-80A. Cessna L-27A **Army model of earlier 310 version.**

A-81. Curtiss O-1   An O-1 Falcon with a Liberty engine.

A-81A. Curtiss O-1E   An improved O-1B Falcon;27 delivered in 1929

A-81B. Curtiss O-1G   The 1931 version of the Falcon; 30 built.

A-81C. Douglas O-2H   The big production version of the O-2 type.

A-81D. Douglas O-7   An O-2 with an inverted Packard engine.

A-82. Douglas O-8 An O-2 with an experimental Curtiss engine.

A-82A. Lockheed XO-10 A 1927 conversion of an XOA-2 amphibian.

A-83. Curtiss XO-13 An O-1 with Conqueror engine for racing.

A-83A. Curtiss XO-13A Another racing Falcon with skin cooling.

A-84. Douglas XO-14 Light observation of 1928 with Whirlwind.

A-85. Curtiss XO-16 An O-11 with Conqueror in place of Liberty.

A-86. Consolidated O-17 A PT-3 redesignated as an observation.

A-86A. Curtiss XO-18 A Falcon testing Curtiss Chieftain engine.

A-87. Thomas Morse O-19B Initial O-19 production type of 1930.

A-88. Thomas Morse O-19C The 1931 version with a tail wheel.

A-88A. Douglas O-22 A 1929 observation version with Wasp engine.

A-89. Douglas O-25A Most-produced O-25 type; Conqueror engine.

A-89A. Douglas O-25B One of three O-25As with dual controls.

A-90. Curtiss O-26 A 1929 Conversion of an O-1E to Conqueror.

A-90A. Fokker YO-27 Three place twin engine service test.

A-91. Vought O-28 Famous Navy O2U- Corsair for Army test.

A-91A. Douglas XO-31 Two high wing monoplanes for 1930 test.

A-91B. Douglas YO-31A Five 1931 service test modifications.

A-91C. Douglas O-32A Thirty Wasp powered planes of 1930.

A-92. Thomas Morse Y1O-33 An O-19B with Conqueror engine, 1930.

A-93. Douglas O-38 Initial design, an O-25A with Hornet engine.

A-93A. Douglas O-38B Most-produced O-38; 63 aircraft in 1931.

A-94. Douglas O-38D One Cyclone powered plane for ANG in 1933.

A-94A. Curtiss Y1O-40B Service test Raven monoplane of 1933.

A-95. Thomas Morse XO-41 Final T-M observation of 1932, no more.

A-96. Douglas XO-46 Final O-43A modified with Twin Wasp Jr.

A-97. Douglas XO-46 Prototype for production O-46A in 1935.

A-97A. Douglas XO-46 Differed from production in aft canopy area.

A-98. No. American XO-47 Single 1936 prototype for O-47A.

A-98A. No. American XO-47   Three place mid wing observation plane.

A-99. Stinson O-49   First light observation type in 1940; to L-1

A-100. Bellanca YO-50   Loser in light observation competition.

A-101. Ryan YO-51   The other losing light observation plane.

A-102. Ryan YO-51   Three aircraft in 1940; no production.

A-103. Stinson YO-54 One of six civil 105s for 1941 tests.

A-104. Douglas OA-4 One of two civil Dolphins; was C-26.

A-104A. Douglas OA-5 Single new amphibian of many designations.

A-167. Curtiss JN-S A "standardized" Jenny trainer.

Consolidated YPT-6 One of ten 1930 service test Fleets.

A-169. Stearman XPT-9 An Army Stearman Cloudboy in 1931.

A-170. Verville Y1PT-10C A 1932 service test trainer.

A-171. Consolidated YPT-12 The final trainer from Consolidated.

A-171A. Waco YPT-14 Army version of UPF-7 in 1940.

A-172. Ryan YPT-16 One of 15 STA trainers of 1940.

A-173. Fairchild PT-19B Aeronca-built version of Fairchild trainer.

A-173A. Fairchild PT-19 Civil rebuild of famous trainer in 1975.

A-173B. Ryan PT-20 1941 Version of PT-16, Men-asco engine.

A-174. Ryan PT-20A PT-20 revised to use Kinner radial in 1941

A-175. Ryan XPT-25 One of five 1942 plywood trainers.

A-174A. Ryan PT-22 Civil rebuild of famous trainer in 1975.

A-175A. Convair T-29B Twin engined Convair at Glen Martin airport.

A-175B. Northrop T-38A Supersonic trainer built in late 1960s.

A-175C. DeHavilland U-1A Famous Army Otter in October, 1961.

A-176. Curtiss X-19 Four tilting propeller V/STOL plane in 1965.

A-177. Aerophysics VZ-7 Flying Jeep by the Curtiss Wright subsidiary